Methods in Molecular Biology

Series Editor
John M. Walker
School of Life and Medical Sciences
University of Hertfordshire
Hatfield, Hertfordshire, UK

For further volumes:
http://www.springer.com/series/7651

For over 35 years, biological scientists have come to rely on the research protocols and methodologies in the critically acclaimed *Methods in Molecular Biology* series. The series was the first to introduce the step-by-step protocols approach that has become the standard in all biomedical protocol publishing. Each protocol is provided in readily-reproducible step-by-step fashion, opening with an introductory overview, a list of the materials and reagents needed to complete . the experiment, and followed by a detailed procedure that is supported with a helpful notes section offering tips and tricks of the trade as well as troubleshooting advice. These hallmark features were introduced by series editor Dr. John Walker and constitute the key ingredient in, each and every volume of the *Methods in Molecular Biology* series. Tested and trusted, comprehensive and reliable, all protocols from the series are indexed in PubMed.

Enteric Glial Cells

Methods and Protocols

Edited by

Giuseppe Esposito

Department of Physiology and Pharmacology "V. Erspamer", Sapienza University of Rome, Rome, Italy

Humana Press

Editor
Giuseppe Esposito
Department of Physiology and Pharmacology "V. Erspamer"
Sapienza University of Rome
Rome, Italy

ISSN 1064-3745 ISSN 1940-6029 (electronic)
Methods in Molecular Biology
ISBN 978-1-0716-4794-3 ISBN 978-1-0716-4795-0 (eBook)
https://doi.org/10.1007/978-1-0716-4795-0

This Humana imprint is published by the registered company Springer Science+Business Media, LLC, part of Springer Nature.
The registered company address is: 1 New York Plaza, New York, NY 10004, U.S.A.

Preface

The enteric nervous system (ENS) is a vast and intricate neural network that governs fundamental gastrointestinal (GI) functions, including motility, secretion, nutrient absorption, and blood flow. It also plays an essential role in processes such as immune and endocrine regulation, host-microbe interactions, and the maintenance of epithelial barrier integrity. This wide-ranging set of homeostatic functions is carried out through a "brain-like" circuitry composed of neurons and glial cells, forming a continuous network that spans the entire digestive tract. The ENS can mediate many of these gut behaviors autonomously from the central nervous system (CNS), although it can act in close collaboration with extrinsic afferent and efferent neurons.

The ENS contains approximately 200 to 600 million neurons, a number comparable to that found in the spinal cord. The number of glial cells within the ENS is even greater—outnumbering neurons by a ratio of 3 to 5. These enteric glial cells and neurons are organized into specialized ganglia, which are interconnected by fibers to form two primary plexuses: the myenteric (Auerbach's) and submucosal (Meissner's) plexuses. The myenteric plexus is located between the longitudinal and circular smooth muscle layers, extending throughout the GI tract from the upper esophagus to the internal anal sphincter. The submucosal plexus is found in the submucosa of the small and large intestines, although it is absent in the esophagus and present only sparsely in the stomach. The myenteric plexus is primarily involved in controlling GI motility, while neurons in both submucosal plexuses essentially modulate the fluid movements, local blood flow, and sensory functions.

Enteric neuronal cell bodies are concentrated in these ganglia and their projections extend extensively throughout the plexuses and to various targets in the mucosa and smooth muscle. Enteric glia, a large and diverse population of glial cells, are closely associated with the enteric neurons, providing crucial support and regulation throughout the digestive tract. These glia share similarities with other types of glia in the nervous system, although studies have shown that enteric glia are unique, with distinct transcriptional profiles that differentiate them from their counterparts in the CNS. The complexity of enteric glia is further revealed by the diversity of their morphological forms and functional roles. Specific molecular markers such as Sox10, S100b, and Plp1 are expressed across all enteric glial subtypes, while Gfap expression is primarily seen in the myenteric glia and fluctuates depending on the state of the tissue, making it a particularly useful marker in studies of intestinal inflammation. This heterogeneity in glial function poses an intriguing question about whether these phenotypic differences reflect variations in response to local stimuli or if distinct types of enteric glia exist with specialized roles.

Recent advances have revealed the bidirectional communication between enteric neurons and glia, which significantly influences the function of the ENS. This emerging perspective is reshaping our understanding of how enteric circuits operate, the regulatory mechanisms involved in maintaining gut health, and how the gut-brain axis contributes to signaling between the gut and brain. Enteric glia are no longer seen as mere passive support cells; rather, they are actively involved in regulating the functions of the ENS, interacting with neurons, immune cells, stem cells, and microbiota.

Furthermore, enteric glia have recently been recognized as key players in gut-brain signaling, potentially influencing brain functions and contributing to neurological disorders.

They also play a pivotal role during intestinal inflammation, acquiring phenotypes that can contribute to diseases such as neuroinflammation, cancer, and infections. Changes in glial networks are observed in various GI diseases, including inflammatory bowel disease (IBD) and irritable bowel syndrome (IBS), as well as in extragastrointestinal conditions. Both losses and gains in glial function can disrupt ENS homeostasis, impacting gut-brain communication and contributing to disease progression.

Despite mounting evidence of the critical roles enteric glia play in health and disease, many fundamental questions remain. Our understanding of the molecular and cellular mechanisms underlying glial function in the ENS is still in its infancy. New genetic tools and research approaches that enable the study of enteric glial cells will be crucial for addressing these gaps.

This book aims to highlight the heterogeneity and plasticity of enteric glial cells and explore their contributions to both GI homeostasis and disease pathophysiology. Through an interdisciplinary approach, this work brings together expertise from neurobiology, immunology, physiology, pharmacology, and pathology to provide a comprehensive perspective on the ENS and its glial components. Electrophysiological techniques have proven invaluable in revealing how enteric glia interact with neurons to influence gut motility and other vital functions. The relationship between enteric glia and the gut microbiota is also a burgeoning area of research, with implications for immune regulation and the maintenance of gut homeostasis.

Advances in immunohistochemistry have allowed researchers to observe morphology, distribution, and functional states of enteric glia in living organisms, providing new insights into their roles in both normal and pathological conditions, including inflammatory bowel diseases. The growing focus on the gut-brain axis has spurred investigations into how enteric glia influence neurodegenerative diseases (such as Parkinson's and Alzheimer's) and mood disorders. These findings reinforce the idea that bridging the gap between GI and neurological functions is essential for understanding the complex relationship between these two systems.

This book represents a collaborative effort from leading researchers around the world, each contributing their expertise to this evolving field. It underscores the importance of cross-disciplinary studies in advancing our understanding of the diverse and critical functions of enteric glial cells. We hope that this work will not only deepen the understanding of enteric glia but also stimulate new research avenues that could lead to breakthroughs in treating a range of gastrointestinal and neurological disorders.

Rome, Italy *Giuseppe Esposito*

Contents

Contributors

LAURA BENVENUTI • *Department of Clinical and Experimental Medicine, University of Pisa, Pisa, Italy*

LOLA BONNEAU • *Center for Stem Cell and Organoid Medicine (CuSTOM), Cincinnati Children's Hospital Medical Center, Cincinnati, OH, USA*

VALENTINA CAPUTI • *Poultry Production and Product Safety Research Unit, Agricultural Research Service, United States Department of Agriculture, Fayetteville, AR, USA*

CARLA CIRILLO • *Toulouse NeuroImaging Center (ToNIC), Inserm U1214, University of Toulouse Paul Sabatier, Toulouse, France*

NINA COLITTI • *Toulouse NeuroImaging Center (ToNIC), Inserm U1214, University of Toulouse Paul Sabatier, Toulouse, France*

VANESSA D'ANTONGIOVANNI • *Department of Clinical and Experimental Medicine, University of Pisa, Pisa, Italy*

PIERLUIGI DE PIERRO • *Department of Clinical Medicine and Surgery, 'Federico II' University of Naples, Naples, Italy*

CLELIA DI SALVO • *Department of Clinical and Experimental Medicine, University of Pisa, Pisa, Italy*

TONY DURAND • *Inserm U1235 TENS, Nantes, France*

GIUSEPPE ESPOSITO • *Department of Physiology and Pharmacology "V. Erspamer", Sapienza University of Rome, Rome, Italy*

MARIA SOLE FACIONI • *Department of Clinical Medicine and Surgery, 'Federico II' University of Naples, Naples, Italy*

SOFIA FAGGIN • *Department of Pharmaceutical and Pharmacological Sciences, University of Padova, Padova, Italy*

FARANAK FATTAHI • *Department of Cellular and Molecular Pharmacology, University of California, San Francisco, San Francisco, CA, USA; Eli and Edythe Broad Center of Regeneration Medicine and Stem Cell Research, University of California, San Francisco, San Francisco, CA, USA; Program in Craniofacial Biology, University of California, San Francisco, San Francisco, CA, USA*

MATTEO FORNAI • *Department of Clinical and Experimental Medicine, University of Pisa, Pisa, Italy*

SILVIA BASILI FRANZIN • *Department of Physiology and Pharmacology "V. Erspamer", Sapienza University of Rome, Rome, Italy*

MARIA CECILIA GIRON • *Department of Pharmaceutical and Pharmacological Sciences, University of Padova, Padova, Italy*

JACQUES GONZALES • *Department of Physiology, Michigan State University, East Lansing, MI, USA*

VLADIMIR GRUBIŠIĆ • *Department of Biomedical Sciences, New York Institute of Technology College of Osteopathic Medicine, Old Westbury, NY, USA; Center for Biomedical Innovation, New York Institute of Technology College of Osteopathic Medicine, Old Westbury, NY, USA*

Brian D. Gulbransen • *Department of Physiology, Michigan State University, East Lansing, MI, USA*
Sirine Hacene • *Toulouse NeuroImaging Center (ToNIC), Inserm U1214, University of Toulouse Paul Sabatier, Toulouse, France*
Austin Irwin • *Department of Biomedical Sciences, New York Institute of Technology College of Osteopathic Medicine, Old Westbury, NY, USA*
Julia Jamka • *Department of Physiology, Michigan State University, East Lansing, MI, USA*
Isabelle Loubinoux • *Toulouse NeuroImaging Center (ToNIC), Inserm U1214, University of Toulouse Paul Sabatier, Toulouse, France*
Elena Lucarini • *Department of Neuroscience, Psychology, Drug Research and Child Health (NEUROFARBA), University of Florence, Florence, Italy*
Maxime M. Mahe • *Université de Nantes, Inserm, TENS, The Enteric Nervous System in Gut and Brain Diseases, IMAD, Nantes, France; Center for Stem Cell and Organoid Medicine (CuSTOM), Cincinnati Children's Hospital Medical Center, Cincinnati, OH, USA; Department of Pediatrics, University of Cincinnati, Cincinnati, OH, USA*
Homa Majd • *Department of Cellular and Molecular Pharmacology, University of California, San Francisco, San Francisco, CA, USA; Eli and Edythe Broad Center of Regeneration Medicine and Stem Cell Research, University of California, San Francisco, San Francisco, CA, USA*
Wilmarie Morales-Soto • *Department of Physiology and Biomedical Engineering, Mayo Clinic, Rochester, MN, USA*
Anoohya N. Muppirala • *Department of Pediatrics, Boston Children's Hospital and Harvard Medical School, Boston, MA, USA*
Irene Palenca • *Department of Physiology and Pharmacology "Vittorio Erspamer", Sapienza University of Rome, Rome, Italy*
Anthony Pellicano • *Department of Biomedical Sciences, New York Institute of Technology College of Osteopathic Medicine, Old Westbury, NY, USA*
Marcella Pesce • *Department of Clinical Medicine and Surgery, 'Federico II' University of Naples, Naples, Italy*
Evgeniia Pimenova • *Department of Cellular and Molecular Pharmacology, University of California, San Francisco, San Francisco, CA, USA*
Meenakshi Rao • *Department of Pediatrics, Boston Children's Hospital and Harvard Medical School, Boston, MA, USA*
Giovanni Sarnelli • *Department of Clinical Medicine and Surgery, 'Federico II' University of Naples, Naples, Italy*
Luisa Seguella • *Department of Physiology and Pharmacology "V. Erspamer", Sapienza University of Rome, Rome, Italy; Department of Physiology, Neuroscience Program, Michigan State University, East Lansing, MI, USA*
Beatriz Thomasi • *Department of Physiology, Michigan State University, East Lansing, MI, USA*
Angela Tietto • *Department of Pharmaceutical and Pharmacological Sciences, University of Padova, Padova, Italy; School of Specialization in Clinical Pharmacology and Toxicology, University of Ferrara, Ferrara, Italy*
Alessandra Toti • *Department of Neuroscience, Psychology, Drug Research and Child Health (NEUROFARBA), University of Florence, Florence, Italy*

Simon Vales • *Université de Nantes, Inserm, TENS, The Enteric Nervous System in Gut and Brain Diseases, IMAD, Nantes, France*
Martina Vincenzi • *Department of Physiology and Pharmacology "Vittorio Erspamer", Sapienza University of Rome, Rome, Italy; Department of Experimental Medicine, University of Rome Tor Vergata, Rome, Italy*
Aurora Zilli • *Department of Physiology and Pharmacology "V. Erspamer", Sapienza University of Rome, Rome, Italy*

Chapter 1

Introducing Enteric Glial Cells

Giuseppe Esposito

Abstract

This book emphasizes the pivotal role of enteric glial cells in maintaining gut health and their influence on the gut–brain axis. Although their existence was first noted at the end of the nineteenth century, recent scientific attention has shed light on their crucial functions, especially in inflammatory bowel diseases and potential links to neurodegenerative conditions like Parkinson's disease. Understanding how these cells operate and interact within the complex network of the enteric nervous system is vital, not only for unraveling gut-related disorders but also for exploring their impact on brain health.

Designed as a comprehensive guide, this book provides practical protocols to identify, visualize, and isolate enteric glial cells using techniques such as enzymatic digestion, immunofluorescence, and advanced imaging methods like confocal microscopy. It highlights the importance of studying glial cell behavior and their morphological diversity across different states of activation and maturation. Moreover, it explores innovative approaches using transgenic models and human biopsies to investigate their activity, interactions with the microbiome, and responses to inflammation.

Understanding these mechanisms may open new avenues for therapies targeting both gastrointestinal and neurodegenerative diseases. This book also discusses possibilities for cell-based interventions, including the potential for enteric glial transplantation into the central nervous system. Overall, this work aims to bridge gastrointestinal and neurological research, underscoring the significance of enteric glia in health, disease, and the intricate communication between the gut and the brain. Advancing knowledge in this area could lead to novel, more precise treatments that address the root causes of complex conditions affecting both the gut and the nervous system.

Key words Enteric Glia, Molecular biology, Multidisciplinary techniques, Neuroscience, Gastroenterology, Pharmacology

Particularly regarding the function of enteric glial cells as major controllers of gut balance, the enteric nervous system (ENS) has become a critical area of study. Although they were first reported at the end of the nineteenth century, enteric glia garnered more attention from scientists in the last few years of the twentieth century [1, 2]. In the context of inflammatory bowel diseases and the possible route for neurodegenerative processes down the gut–brain axis, these cells have major functions [3, 4]. Understanding the basic mechanisms controlling enteric glial cells becomes vital as

Giuseppe Esposito (ed.), *Enteric Glial Cells: Methods and Protocols*, Methods in Molecular Biology, vol. 2971, https://doi.org/10.1007/978-1-0716-4795-0_1,

scientific interest in the interaction of the gut and the brain keeps rising.

This book appeals to a wide range of interdisciplinary scientists since it contains a thorough set of procedures meant to simplify biomolecular studies of enteric glial cells. From physiologists looking at gut function to pharmacologists researching potential therapeutic targets and neuroscientists studying the neural bases of gut–brain communication, this should be a helpful resource throughout many disciplines. Examining key methodological approaches for spotting, depicting, and isolating glial cells in both the enteric nervous system and isolated settings, the chapters offer a detailed overview of techniques and their experimental applications. By various methods, including isolation from transgenic animals, intestinal organoids, electrophysiology studies, calcium imaging, and transfection techniques, we will investigate several functional research techniques for enteric glia. These approaches not only deepen our knowledge of glial cell biology but also offer opportunities for fresh treatments.

The primary goal of this book is to provide a clear and practical guide to essential techniques for identifying, visualizing, and isolating enteric glial cells, both within the enteric nervous system and in isolated laboratory conditions. These methods include isolating enteric glia through enzymatic digestion, which can be performed on samples obtained from transgenic animal models or human biopsies. We also explore techniques for visualizing and characterizing these cells using immunofluorescence staining with specific markers such as glial fibrillary acidic protein (GFAP) and S100 calcium-binding protein B (S100B). Additionally, our goal is to define the uniqueness of this cell population and to elucidate the morphological diversity associated with different activation and maturation states. In this context, we employ a more in-depth methodological approach to evaluate Sox10 and Plp1 expression depending on the reactivity or quiescent state of enteric glia under various conditions. Using confocal microscopy, we aim to reveal detailed structural features and the connectivity of glial networks, enhancing our understanding of their functional complexity.

Additionally, this book explores how to study the behavior and functions of enteric glia by utilizing transgenic mice that express fluorescent proteins, allowing researchers to observe live glial activity within intact tissues. These techniques are designed to bring greater understanding of the vital roles that enteric glial cells play in gut health and disease. Their position as a vital link in the neuro-immunological interface becomes more obvious as we gain more understanding of enteric glial cells [5]. These cells are not only involved in the neural architecture of the gut; they also help to moderate the impact of disturbances in the gut environment that could cause changes in central nervous system (CNS) activity [5–7]. Studies have started to show how gut inflammatory processes can spiral out of control on brain health, hence affecting neuropsychiatric and neurological conditions. Therefore, the enteric glia acts as a solitary "*trait d' union*" between intestinal neuropathology and its central neurological consequences.

Furthermore, we provide guidelines for isolating enteric glial cells from human biopsy samples, therefore permitting study of the interactions of glial cells and the gut microbiota [8, 9]. Deciphering sophisticated signaling pathways in the gut relies on knowing these interactions and emphasizes how microbiome changes affect gut health and general disease conditions. The knowledge from these papers could have far-reaching results, indicating that improving glial health could relieve symptoms in several neuropsychiatric disorders impacted by gut dysbiosis [10–13].

Furthermore, we will describe techniques for isolating and evaluating the activity of enteric glial cells under normal and pathological circumstances, thereby giving knowledge on their functional roles in maintaining balance [14, 15] and responding to different insults [4, 14, 16]. It is critical in this setting to grasp how enteric glial cells react to inflammatory signals and how their interaction with neurons and immune cells is affected [17]. Lastly, considering their autologous transplantable type [18, 19], we will go over creative techniques for enteric glial cell isolation and autotransplanting in the central nervous system [20, 21]. These developments could give fascinating treatments for tackling neurological conditions, possibly originating from or aggravated by gut malfunction. By means of these thorough strategies, this book seeks better knowledge of enteric glial cells and their consequences in gut–brain communication, thereby ultimately closing the distance between gastrointestinal and neurological research in the larger discipline of neurobiology.

Studying the enteric glia today is not just about understanding their role in initiating and sustaining acute and chronic inflammatory processes in intestinal diseases such as ulcerative colitis or Crohn's disease. It is also essential for uncovering new players involved in the origin of primary alterations in the enteric nervous system, which may eventually contribute to the development of neurodegenerative diseases like Parkinson's and Alzheimer's disease, along the gut–brain axis [22]. This research is therefore crucial for identifying new, more precise, and reliable therapeutic targets that could lead to improved treatments for both intestinal and neurodegenerative conditions. Encouragement of more study into this vital field by means of a more thorough understanding of the many roles of the enteric glia will pave the way for creative therapies targeting the complex linkages between gut health and brain function and help to advance the ongoing debate on the importance of the gut–brain axis in neuromodulation and neuroprotection.

In the interest of fostering collaborative research on enteroglial cells across different fields, I want to thank Dr. Luisa Seguella for her support and helpful insights regarding the techniques described in this book, as well as all the friends and colleagues who participated in this project and to the steadily increasing number of researchers who enter the field.

References

1. Santhosh S et al (2024) From diversity to disease: unravelling the role of enteric glial cells. Front Immunol 15:1408744. https://doi.org/10.3389/fimmu.2024.1408744
2. Coelho-Aguiar Jde M et al (2015) The enteric glia: identity and functions. Glia 63(6):921–935. https://doi.org/10.1002/glia.22795
3. Seguella L et al (2019) Play in advance against neurodegeneration: exploring enteric glial cells in gut-brain axis during neurodegenerative diseases. Expert Rev Clin Pharmacol 12(6): 555–564. https://doi.org/10.1080/17512433.2019.1612744
4. Seguella L, Gulbransen BD (2021) Enteric glial biology, intercellular signalling and roles in gastrointestinal disease. Nat Rev Gastroenterol Hepatol 18(8):571–587. https://doi.org/10.1038/s41575-021-00423-7
5. Progatzky F et al (2021) Regulation of intestinal immunity and tissue repair by enteric glia. Nature 599(7883):125–130. https://doi.org/10.1038/s41586-021-04006-z
6. Esposito G et al (2017) HIV-1 Tat-induced diarrhea evokes an enteric glia-dependent neuroinflammatory response in the central nervous system. Sci Rep 7(1):7735. https://doi.org/10.1038/s41598-017-05245-9
7. D'Antongiovanni V et al (2024) Enteric glia and brain astroglia: complex communication in health and disease along the gut-brain axis. Neuroscientist 30(4):493–510. https://doi.org/10.1177/10738584231163460
8. Kato R et al (2024) Indigenous gut microbiota constitutively drive release of ciliary neurotrophic factor from mucosal enteric glia to maintain the homeostasis of enteric neural circuits. Front Immunol 13(15):1372670. https://doi.org/10.3389/fimmu.2024.1372670
9. Seguella L et al (2023) Mini-review: Interaction between intestinal microbes and enteric glia in health and disease. Neurosci Lett 29(806):137221. https://doi.org/10.1016/j.neulet.2023.137221
10. Linan-Rico A (2023) Mini-review: Enteric glial cell reactions to inflammation and potential therapeutic implications for GI diseases, motility disorders, and abdominal pain. Neurosci Lett 812:137395. https://doi.org/10.1016/j.neulet.2023.137395
11. Le Berre C et al (2023) Enteric glia at center stage of inflammatory bowel disease. Neurosci Lett 13(809):137315. https://doi.org/10.1016/j.neulet.2023.137315
12. Yang H et al (2022) Intestinal permeability, dysbiosis, inflammation and enteric glia cells: the intestinal etiology of Parkinson's disease. Aging Dis 13(5):1381–1390. https://doi.org/10.14336/AD.2022.01281
13. Seguella L et al (2020) Leaky gut, dysbiosis, and enteric glia activation: the trilogy behind the intestinal origin of Parkinson's disease. Neural Regen Res 15(6):1037–1038. https://doi.org/10.4103/1673-5374.270308. PMID: 31823880; PMCID: PMC7034261
14. Cabarrocas J et al (2003) Role of enteric glial cells in inflammatory bowel disease. Glia 41(1): 81–93. https://doi.org/10.1002/glia.10169
15. Brown RM et al (2025) Functional analysis of antigen presentation by enteric glial cells during intestinal inflammation. Glia 73(2): 291–308. https://doi.org/10.1002/glia.24632
16. Laddach A et al (2023) A branching model of lineage differentiation underpinning the neurogenic potential of enteric glia. Nat Commun 14(1):5904. https://doi.org/10.1038/s41467-023-41492-3
17. Gonzales J, Gulbransen BD (2025) The physiology of enteric glia. Annu Rev Physiol 87(1): 353–380. https://doi.org/10.1146/annurev-physiol-022724-105016
18. Esposito G et al (2016) Autologous transplantation of intestine-isolated glia cells improves neuropathology and restores cognitive deficits in β amyloid-induced neurodegeneration. Sci Rep 4(6):22605. https://doi.org/10.1038/srep22605

19. Sunardi M, Cirillo C (2023) Mini-review: "Enteric glia functions in nervous tissue repair: therapeutic target or tool?". Neurosci Lett 812:137360. https://doi.org/10.1016/j.neulet.2023.137360
20. Hansebout CR et al (2012) Enteric glia mediate neuronal outgrowth through release of neurotrophic factors. Neural Regen Res 7(28):2165–2175. https://doi.org/10.3969/j.issn.1673-5374.2012.028.001
21. Jiang S et al (2003) Enteric glia promote regeneration of transected dorsal root axons into spinal cord of adult rats. Exp Neurol 181(1):79–83. https://doi.org/10.1016/s0014-4886(02)00030-4
22. D'Antongiovanni V et al (2023) Enteric glia and brain astroglia: complex communication in health and disease along the gut-brain axis. Neuroscientist 30(4):493–510. https://doi.org/10.1177/10738584231163460

Chapter 2

Immunohistochemical Localization and Anatomical Distribution of Enteric Glia Cells in Enteric Nervous System

Sofia Faggin, Valentina Caputi, Angela Tietto, and Maria Cecilia Giron

Abstract

The enteric nervous system (ENS) comprises highly heterogeneous glia cell populations, which play critical roles in the regulation of intestinal neuromuscular function, providing physical and immune protection to neuronal cells and processes in physiological and pathological conditions. Due to the high muscle/mucosa content, thinness, and softness, freezing and sectioning of intestinal tissue are difficult, and distortion of tissue morphology might occur. However, whole-mount immunostaining for glia characterization in the 3D environment of the ENS is a better choice, as this powerful technique facilitates the visualization of different cell populations and their interaction with other cells within the neuromuscular compartments of the intestinal wall, without tissue distortion or damage. In the last decades, novel methodologies for whole-mount staining have been described to preserve the structural integrity of all the neuroglial cell units across the intestinal wall, allowing high-resolution image acquisition. In this chapter, we describe a step-by-step protocol to perform immunofluorescence staining of enteric glial cells in both the myenteric and submucosal plexuses of different regions of the gastrointestinal tract and to optimally visualize cellular constituents of the ENS with confocal microscopy.

Key words Enteric glial cells, Glial markers, Antibodies, Myenteric plexus, Submucosal plexus, Whole mount, Immunohistochemistry, Confocal microscopy, Clarity

1 Introduction

Enteric glia cells (EGCs) belong to the large population of non-myelinating peripheral neuroglia that derives from neural crest precursors and represents one of the components of the enteric nervous system (ENS). Depending on the species and the region of the gastrointestinal tract, EGCs are slightly more numerous than enteric neurons, embracing them in ganglia, a dense network of connections, along the wall of the digestive tract, to

Authors Sofia Faggin and Valentina Caputi have equally contributed to this chapter.

Supplementary Information The online version contains supplementary material available at https://doi.org/10.1007/978-1-0716-4795-0_2.

Giuseppe Esposito (ed.), *Enteric Glial Cells: Methods and Protocols*, Methods in Molecular Biology, vol. 2971,
https://doi.org/10.1007/978-1-0716-4795-0_2,

form the myenteric and submucosal plexus, which are placed between the longitudinal and circular muscle layers and in the submucosa, respectively [1]. These cells were originally considered to be a passive structural support for enteric neurons; however, in the last years, they have emerged to act as cellular integrators in the control of gut motility reflexes, barrier function, and inflammation [1–5].

Mature enteric glial cells display morphological similarity to astrocytes, confirmed by the expression of similar biomarkers such as the glial fibrillary acidic protein (GFAP) [6], the filament protein vimentin [7], the gap junctional channel connexin-43 [8], and S100 calcium-binding protein β (S100β) [9, 10]. However, no exclusive distinctive marker of EGCs has been so far identified since the transcriptional profile of EGCs results markedly to overlap those of oligodendrocytes (e.g., myelin proteolipid protein 1, PLP-1 [11, 12]), astrocytes, and even neurons (e.g., SOX-10) [13] of the central nervous system to highlight their significant plasticity and heterogeneity [1].

To characterize regional heterogeneity and plasticity between the enteric glia population and its association with other cells and tissues during both steady-state and pathological conditions, several techniques can be applied [14], among which immunofluorescence (IF) staining is well suited for high-resolution, temporo-spatial analysis. Based on the specificity of antibody–antigen recognition, IF labeling offers the opportunity to picture different molecular, cellular, or acellular structures in a variety of samples, including cell lines, tissue sections, and whole-mount preparations. This powerful but rather straightforward method may uncover the key players of biomolecular processes, thus enabling scientists to comprehend biological events in vivo, ex vivo, and in vitro. This immunochemical technique depends on the high affinity and specificity of an antibody to interact with a specific epitope on a given antigen. To acquire a strong and specific signal for IF imaging, the choice of primary and secondary antibodies (or antibody mix), as well as the methodology of tissue preparation and tissue fixation, is essential. The intestinal wall has a complex anatomical architecture comprising several layers of different tissues, such as smooth muscles, the enteric nervous system, and the epithelial mucosal barrier. Their intercellular interactions are challenging to map using classical section-based or whole-mount histology. In these last 10 years, several methods and reagents have been proposed to assist in whole tissue processing and subsequent 3D imaging. The introduction of optical tissue clearing to render biological specimens transparent has led to the development of a variety of protocols aimed at detailed high-resolution 3D imaging of the intact whole tissue and organs by confocal, light-sheet, or two-photon microscopy. This technique permits the removal of light-scattering and/or opaque components (e.g., lipids) by unifying the refractive indices throughout the whole biological specimen [15–20]. The mapping of multiple biomarkers' distribution in diverse cell populations or subpopulations and related spatial interactions throughout the whole tissue can be achieved in 3D by combining tissue clearing with IF staining. The major challenges in the immunostaining phase are (1) defining antibodies suitability for the tissue clearing technique and (2) achieving a profound and uniform stain penetration throughout the thickness of the tissue sample. Like other immunoassays (e.g., Western blots, flow cytometry, etc.), the tissue clearing methodology requires antibody validation for 3D IF since the tissue sizes are far larger compared to traditional immunohistochemistry biological sample sections, thus requiring a higher concentration of antibodies, which needs to be optimized.

Here, we describe a straightforward, low-toxicity, time-efficient, and inexpensive tissue clearing protocol for full-thickness 3D IF staining of the gastrointestinal wall. The use of "clearing-enhanced 3D" (Ce3D) technique offers the opportunity to simultaneously achieve outstanding tissue transparency, preservation of fluorescence from diverse reporter proteins and organic fluorophores, maintenance of cellular/tissue architecture, and a capacity for highly multiplexed antibody-based immunolabeling.

Here, we applied Ce3D to confocal imaging of EGCs and neurons. The high signal quality and spatial resolution achieved with Ce3D permit accurate quantitative analysis of ENS phenotypic profiles in intestinal tissues. This protocol preserves the integrity of both the submucosal plexus and myenteric plexus of the intestinal wall. The experimental procedure takes 4 days and was adapted from the methodologies previously published [21, 22].

2 Materials

Prepare and store all reagents and solutions at 4 °C (unless indicated otherwise). Follow all waste disposal regulations when disposing of waste materials.

2.1 Sample Preparation and Fixation

1. Chemical fume hood.
2. Fine scissors, fine tissue tweezers, surgical hemostats with blunt ends, and a scalpel blade.
3. 50 mL conical Falcon tubes.
4. 2 mL Eppendorf tubes.
5. Tube racks.
6. Cotton threads.
7. 2 mL syringe.
8. Button needle.

Table 1
List of substances required to perform the fixation and immunostaining

Substances
Paraformaldehyde 8% aqueous solution
Thimerosal (sodium merthiolate)
1-Thioglycerol
Bovine serum albumin (BSA)
Histodenz
N-methyl acetamide
Normal horse serum (NHS)
Triton X-100

9. Orbital shaker.
10. Pasteur pipettes.
11. Phosphate buffered saline (PBS): Dilute 10× PBS in Milli Q water to obtain 1× PBS (*see* **Note 1**).
12. 4% Paraformaldehyde (PFA) solution: In a chemical fume hood, for a total of 20 mL of 4% PFA, add 10 mL vial of 8% PFA to the conical tube containing 2 mL of 10× PBS + 8 mL of Milli Q water (*see* **Note 2** and Table 1).
13. 0.05% Thimerosal (sodium merthiolate) solution: In a chemical fume hood, for a total of 45 mL of 0.05% sodium merthiolate, dissolve 22.5 mg Thimerosal in 45 mL of 1× PBS (Table 1).

2.2 Mucosa Scraping

1. Stereomicroscope.
2. Fine tissue tweezers and a scalpel blade.
3. Rubber dish.
4. Pasteur pipettes.
5. 48-well plate.
6. Phosphate buffered saline (PBS): *see* Subheading 2.1, **step 10** (*see* **Note 1**).

2.3 Blocking, Permeabilization, and Staining

1. Fine tissue tweezers.
2. Orbital shaker.
3. 10-, 200-, and 1000-μL micropipettes and corresponding tips.
4. 48-well plates.
5. Pasteur pipettes.

Table 2
Characteristics of primary and secondary antibodies for enteric glial cells labeling in immunofluorescence analysis

Primary antisera				
Antibody	Host species	Clone	Concentration/amount	Dilution
HuC/D	Mouse biotin-conjugated	Monoclonal	5–20 μg/mL	1:100
SOX10	Rabbit	Monoclonal	0.1–0.2 mg/mL	1:100
GFAP	Chicken	Polyclonal	1 mg/mL	1:100
S100β	Guinea pig	Polyclonal	100 μL	1:100
PLP1	Rabbit	Polyclonal	1 mg/mL	1:50
Secondary antisera				
Antibody		Clone	Concentration	Dilution
Goat anti-rabbit IgG Alexa 488-conjugated (H + L)		Polyclonal	2 mg/mL	1:1000
Goat anti-Guinea pig IgG Alexa 488-conjugated (H + L)		Polyclonal	2 mg/mL	1:1000
Goat anti-chicken IgY Alexa 488 conjugated (H + L)		Polyclonal	2 mg/mL	1:1000
Streptavidin Alexa 555-conjugated		–	2 mg/mL	1:1000

6. Blocking solution (5% v/v Normal Horse Serum (NHS), 5% w/v bovine serum albumin (BSA), and 0.5% v/v Triton X-100 in 1× PBS): For a total of 100 mL of blocking solution, add 5 mL of NHS, 5 g BSA, and 0.5 mL of Triton X-100 in 1× PBS (*see* **Note 3** and Table 1).
7. Ce3D washing solution (0.05% v/v Triton X-100, 0.5% v/v 1-thioglycerol in 1× PBS): For a total of 100 mL of Ce3D washing solution, add 0.05 mL of Triton X-100 and 0.5 mL of 1-thioglycerol in 1× PBS (*see* **Note 4** and Table 1).
8. Primary antibody solution: Dilute the chosen primary antibody in blocking solution (*see* **Note 5**, Table 2). Use the recommended antibody dilution for immunofluorescence staining.
9. Secondary antibody solution: In case of performing indirect immunofluorescence using an unconjugated primary antibody followed by the incubation of a labeled secondary antibody, dilute the labeled secondary antibody of choice in blocking solution (*see* **Notes 5** and **6**, Table 2). Use the recommended secondary antibody dilution for immunofluorescence staining.

2.4 Tissue Clearing Medium

1. Fine tissue tweezers.
2. 10-, 200-, and 1000-μL micropipettes and corresponding tips.
3. Non-treated, sterilized 48-well tissue culture plates.

4. 0.5 mL Eppendorf tubes.
5. Parafilm.
6. Orbital shaking incubator.
7. Electronic serological pipettor.
8. 10 mL serological pipettes.
9. Glass bottle (100 mL).
10. Magnetic bar.
11. Magnetic stirrer and heater.
12. 40% v/v N-methyl acetamide solution (Table 1):
    - Place the glass bottle containing N-methylacetamide (100%) and a 10 mL serological pipette into the incubator shaker at 37 °C until fully melted (*see* **Note 7**).
    - Transfer 10 mL of N-methylcetamide into a 50 mL conical tube using the 10 mL prewarmed serological pipette inserted in the automatic pipettor (*see* **Note 8**).
    - Add 15 mL of 1× PBS at room temperature (RT) to 10 mL of N-methylacetamide and mix to obtain 40% (vol/vol) N-methylacetamide stock solution (*see* **Note 9**).
13. Ce3D medium preparation: Prepare 30 mL of this solution using a clean glass bottle and a clean magnetic bar. Place the bottle on the magnetic stirrer and heater. Prepare the Ce3D medium solution by adding the following chemicals in the following sequential order:
    1. 12 mL of 40% N-methylacetamide solution (*see* **Note 10**).
    2. 24 g of Histodenz (Table 1).
    3. 4.5 mL of 40% N-methylacetamide solution (*see* **Note 10**).
    4. 0.03 mL of Triton X-100 (0.1% v/v) (*see* **Note 4**).
    5. 0.15 mL of 1-thioglycerol (0.5% v/v) (*see* **Note 4**).
14. Heat the solution at 40–60 °C while stirring until completely dissolved to obtain a transparent solution, ensuring that the solution does not boil. Seal the bottle with parafilm and store it at RT.

2.5 Sample Mounting

1. Fine tissue tweezers.
2. Stereomicroscope.
3. Glass cover slips #1.5 (0.16–0.19 mm thickness).
4. Pasteur pipettes.
5. Confocal microscope.

3 Methods

Make sure to follow all laboratory safety regulations when preparing and performing the experiment. Perform all procedures at RT, unless otherwise specified.

3.1 Preparation for Tissue Harvesting

Make sure that the PFA 4% solution is completely thawed and totally transparent. Always use freshly thawed aliquots. For optimal fixation, keep PFA 4% at 4 °C.

3.2 Tissue Harvesting

Sections of the ileum/colon 3–5 cm in length are gently rinsed with a 2 mL button-needle syringe filled with 1× PBS and placed in a glass beaker containing PBS 1X.

3.3 Tissue Fixation and Whole-Mount Preparation

1. Prepare two 10 cm cotton threads.
2. Seal the bottom end of the intestinal segment, making a double knot with one 10 cm cotton thread (*see* **Note 11**).
3. On the one hand, handle the intestinal segment from the top end to expose the open part of the tube and flush 4% PFA with a 2 mL button-needle syringe until the intestinal tube appears turgid and well-stretched (*see* **Note 12** and Fig. 1).
4. Clamp the open top end of the intestinal segment with a surgical hemostat to prevent leakage of 4% PFA.
5. Make a double knot with a cotton thread below the hemostat to seal the top end of the intestinal segment.
6. Remove the surgical hemostat to obtain a "sausage-like" intestinal preparation (*see* Fig. 1).
7. Place the intestinal sausage-like preparation in a 50 mL conical tube prefilled with ~10 mL of 4% PFA (*see* **Note 13**) and place it at 4 °C (wet ice) for 3 h.

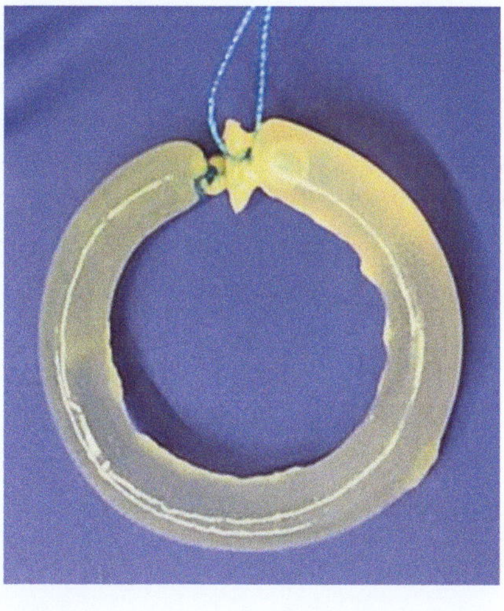

Fig. 1 Sausage-like intestinal preparation filled with 4% PFA

8. In the chemical fume hood, with the help of scissors, cut off the sealed ends of the intestinal sausage-like preparation to flow out the PFA.
9. Place the opened intestinal sausage-like preparation in a conical tube containing ~20 mL of 1× PBS to wash away the remaining PFA, and place the tube in the orbital shaker for 30 min at RT (*see* **Note 14**). At the end of the washing step, discard the PBS and fill the same tube with fresh 1× PBS (20 mL). Repeat the washing step for a total of three 30-min washes with 20 mL of 1× PBS each.
10. After finishing the three washes, in the chemical fume hood, place the tissue (from now on denominated whole-mount preparation) into a 2 mL tube containing 0.05% sodium merthiolate solution (*see* **Notes 15** and **16**) and store it at 4 °C.

3.4 Immunostaining

The following steps are usually performed over 4 days.

3.4.1 Tissue Dissection (Day 1)

1. Under a dissecting stereomicroscope, place the sausage-like intestinal preparation onto a rubber dish containing 100 μL of 1× PBS (*see* **Note 17**).
2. Cut each intestinal preparation into segments of ~0.5 cm in length.
3. Open each segment along the mesenteric line to expose the mucosal layer facing the objective for visualizing the villi.
4. Firmly block the tissue with a tweezer and gently scrape the villi with a scalpel blade or glass coverslip (depending upon the age of the experimental animal), being careful not to damage the underlying tissue layers.
5. Transfer the dissected tissues to a 48-well plate containing 100 μL of 1× PBS/well, where each whole-mount tissue will be free-floating.

3.4.2 Blocking and Permeabilization

Transfer the intestinal whole mounts to new clean wells prefilled with 150 μL of blocking solution and place the plate on an orbital shaker for 24 h (overnight; gentle shaking, room temperature).

3.4.3 Staining with Primary Antibodies (Day 2)

1. Remove the blocking solution from the wells and replace it with 100 μL of the appropriately diluted primary antibody solution.
2. Incubate the plate for 24 h (overnight) at RT on an orbital shaker (gentle shaking is applied for at least for 2–3 h, then the plate is kept on a flat surface of the bench, covered in aluminum foil without shaking for the whole time of incubation of the primary antibody).

3.4.4 Washing Step and Staining with Secondary Antibodies (Day 3)

1. Transfer the intestinal whole mounts in new clean wells, prefilled with 100 μL of Ce3D washing solution/well, to wash away the primary antibody solution.
2. Leave the plate on an orbital shaker for 10 min (moderate shaking, RT) and repeat three times (10 min each).
3. After the last washing step, transfer the intestinal whole mounts to new clean wells prefilled with 100 μL/well of the appropriately diluted secondary antibody solution.
4. Incubate the plate for 24 h (overnight) at RT on an orbital shaker (gentle shaking is applied for at least for 2–3 h, then the plate is kept on a flat surface of the bench, covered in aluminum foil without shaking for the whole time of incubation of the secondary antibody).

3.4.5 Washing Step and Tissue Clearing (Day 4)

1. Transfer the intestinal whole mounts in new clean wells, prefilled with 100 μL of Ce3D washing solution/well, to wash away the secondary antibody solution.
2. Leave the plate on an orbital shaker for 10 min and repeat three times (10 min each) (*see* **Note 18**).
3. After the washing steps, dry the intestinal whole mounts on a clean paper towel.
4. Place the intestinal whole mounts in new Eppendorf tubes prefilled with clearing Ce3D medium (usually a minimum of 300 mL/tube).
5. Seal tube with parafilm and incubate at RT for a minimum of 24 h in the dark until imaging step (*see* **Note 19**). Tissues can be stored in the Ce3D medium at RT in the dark, sealed, and the immunofluorescence is stable up to 6 months.

3.5 Mounting

1. Using the stereomicroscope, with fine forceps, gently place the cleared whole-mount tissue on a glass coverslip (#1.5); carefully spread the tissue on the slide without breaking it. If necessary, use a few drops of Ce3D medium solution to facilitate this step (*see* **Note 20**).
2. Carefully place another identical glass coverslip (#1.5) on top of the tissue sample, avoiding bubbles.
3. Wait at least 1 h and preferably overnight before imaging to ensure that the Ce3D medium is evenly distributed throughout the sample.
4. Whole-mount tissue samples are ready for image acquisition (confocal microscopy).

3.6 Image Acquisition and Analysis

A detailed description of imaging techniques is beyond the scope of this chapter; imaging can be undertaken on a spinning disk or laser scanning confocal microscope. Raw images are exported in the most suitable format (e.g., CZI, LIF, etc.) and processed using a variety of software, such as Imaris (Oxford Instruments) or ImageJ (NIH). Examples of the images and videos that can be obtained with this technique are presented in Figs. 2 and 3 and Electronic Supplementary Video 1.

Fig. 2 Representative microphotographs showing the distribution of the pan neuronal marker HuC/D (cyan), and the ECG markers GFAP (green) and S100β (magenta) in the myenteric plexus (left panel) and submucosal plexus (right panel) within the same tissue specimen of mouse colon. Tissues were subjected to immunohistochemistry and tissue clearing with Ce3D solution. Images were acquired with a Stellaris 5 confocal microscope (Leica). Scale bars = 50 μm

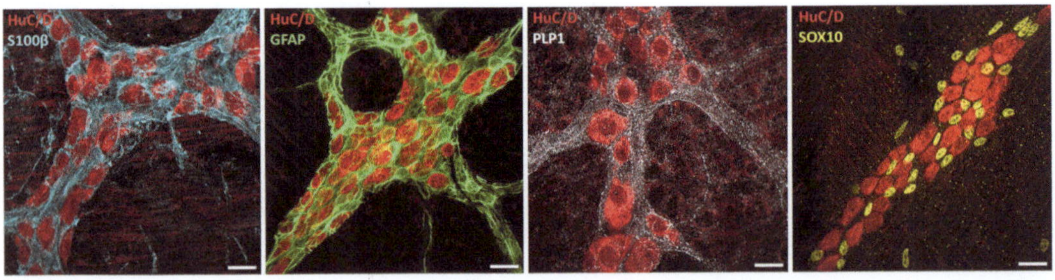

Fig. 3 Whole-mount double immunofluorescence staining of EGCs in the mouse ileum myenteric plexus. Small intestine specimens were stained using primary antibodies to the pan-neuron protein HuC/D (red; **a–d**) and EGCs markers S100β (cyan), GFAP (green), PLP1 (white), and SOX10 (yellow). EGCs staining with S100β (cyan), GFAP (green), and PLP1 (white) reveals that these cells surround enteric neurons, whereas the neural crest marker SOX10 marks all EGCs nuclei (yellow). Tissues were subjected to immunohistochemistry and tissue clearing with Ce3D solution. Images were acquired with a Zeiss LSM 800 confocal microscope. Scale bars = 22 μm

4 Notes

1. Make sure the pH of PBS is 7.4 and adjust if required.
2. To optimize staining, it is critical to use fresh 4% PFA solution, prepared in a chemical fume hood with appropriate personal protective equipment.
3. BSA powder, when added to the solution, can be very sticky. Make sure the BSA is completely dissolved before continuing. If necessary, leave the PBS/BSA solution shaking overnight at 4 °C.
4. Due to the high viscosity of Triton X-100 and 1-thioglycerol, it is helpful to cut the end of a plastic micropipette tip with scissors to be able to properly pipette the required volume of Triton X-100 or 1-thioglycerol in 1× PBS. CAUTION: 1-thioglycerol is a skin irritant.
5. There is no need to wash the blocking solution away as the primary antibody is diluted in blocking solution. When selecting the primary antibody, it is recommended that the species in which the antibody was generated is different from that of the assay sample to avoid cross-reactivity of the secondary antibody with the endogenous immunoglobulins. Furthermore, be careful to use primary antibodies raised in different species when targeting more than one antigen, so that the diverse fluorescently conjugated secondary antibodies used in the next step identify exclusively one species.
6. Cover the plate with aluminum foil to protect the sample from the light. When selecting the secondary antibody, it is recommended that the species in which the antibody was generated is the same as that of the previously incubated primary antibody. Furthermore, if more than one antigen has to be examined, be careful to select fluorescence-conjugated secondary antibodies that emit at different wavelengths.
7. CAUTION: N-methylacetamide is supposed to be a human reproductive and fetal toxicant. All steps must be performed with appropriate lab safety equipment and appropriate personal protective equipment, and when open to air, must be done in a chemical fume hood until mounted and sealed. N-methylacetamide melts at 28.0 °C; therefore, place the container in an incubation shaker. When N-methylacetamide melts, work quickly in the chemical fume hood, as it will solidify in a few minutes. Collect waste in appropriate containers for disposal.
8. N-methylacetamide will stick to the pipette wall if the pipette is not prewarmed.

9. The 40% N-methylacetamide (v/v) can be stored tightly capped with parafilm at RT for 2 months.
10. Use a second clean and prewarmed serological pipette to add the required amount of N-methylacetamide.
11. Please ensure to securely close the end of the intestine with the thread and, if needed, make a third knot.
12. Filling the intestine with PFA 4% is a critical step, since the overfilling can result in the breaking up of the intestinal tissue, whereas underfilling can compromise the subsequent immunohistochemical analysis.
13. It is important to fully submerge the intestinal sausage-like preparation in 4% PFA solution to ensure a homogenous fixation.
14. It is crucial to wash the samples with freshly prepared 1× PBS at least three times in a chemical fume hood.
15. CAUTION: The preservative and antimicrobial agent sodium merthiolate contains mercury; therefore, it must be handled in a chemical fume hood with appropriate personal protective equipment.
16. The opened and washed intestinal sausage-like preparation should be completely submerged in the sodium merthiolate solution and fully distended with no folds.
17. Harvest the intestinal preparation, clamping only the mesentery to avoid any damage to the intestinal tissue.
18. If needed, classic fluorescent nuclear and chromosome counterstain (e.g., DAPI, TOTO-3, etc.) can be added at this step of the procedure. After incubation, make sure to perform five 10-min washes to remove the dye.
19. It is recommended to image the samples 24–48 h after mounting.
20. Tissues are imaged while mounted in the Ce3D medium; no Mowiol mounting medium is required.

References

1. Grubišić V, Gulbransen BD (2017) Enteric glia: the most alimentary of all glia. J Physiol 595(2):557–570. https://doi.org/10.1113/jp271021
2. Linan-Rico A, Ochoa-Cortes F, Schneider R, Christofi FL (2023) Mini-review: enteric glial cell reactions to inflammation and potential therapeutic implications for GI diseases, motility disorders, and abdominal pain. Neurosci Lett 812:137395. https://doi.org/10.1016/j.neulet.2023.137395
3. Progatzky F et al (2021) Regulation of intestinal immunity and tissue repair by enteric glia. Nature 599(7883):125–130. https://doi.org/10.1038/s41586-021-04006-z
4. Seguella L, Gulbransen BD (2021) Enteric glial biology, intercellular signalling and roles in gastrointestinal disease. Nat Rev Gastroenterol Hepatol 18(8):571–587. https://doi.org/10.1038/s41575-021-00423-7
5. Sharkey KA (2015) Emerging roles for enteric glia in gastrointestinal disorders. J Clin Invest

125(3):918–925. https://doi.org/10.1172/jci76303

6. Jessen KR, Saffrey MJ, Burnstock G (1983) The enteric nervous system in tissue culture. I. Cell types and their interactions in explants of the myenteric and submucous plexuses from Guinea pig, rabbit and rat. Brain Res 262(1):17–35. https://doi.org/10.1016/0006-8993(83)90466-3
7. Jessen KR, Mirsky R (1983) Astrocyte-like glia in the peripheral nervous system: an immunohistochemical study of enteric glia. J Neurosci 3(11):2206–2218. https://doi.org/10.1523/jneurosci.03-11-02206.1983
8. McClain J et al (2014) Ca2+ responses in enteric glia are mediated by connexin-43 hemichannels and modulate colonic transit in mice. Gastroenterology 146(2):497–507.e491. https://doi.org/10.1053/j.gastro.2013.10.061
9. Cerantola S et al (2020) Involvement of enteric glia in small intestine neuromuscular dysfunction of toll-like receptor 4-deficient mice. Cells 9(4). https://doi.org/10.3390/cells9040838
10. Marsilio I et al (2021) Oxidized phospholipids affect small intestine neuromuscular transmission and serotonergic pathways in juvenile mice. Neurogastroenterol Motil 33(4):e14036. https://doi.org/10.1111/nmo.14036
11. Rao M et al (2015) Enteric glia express proteolipid protein 1 and are a transcriptionally unique population of glia in the mammalian nervous system. Glia 63(11):2040–2057. https://doi.org/10.1002/glia.22876
12. Boesmans W et al (2022) Development, diversity, and neurogenic capacity of enteric glia. Front Cell Dev Biol 9. https://doi.org/10.3389/fcell.2021.775102
13. Young HM, Bergner AJ, Müller T (2003) Acquisition of neuronal and glial markers by neural crest-derived cells in the mouse intestine. J Comp Neurol 456(1):1–11. https://doi.org/10.1002/cne.10448
14. Fu YY et al (2013) 3-D imaging and illustration of mouse intestinal neurovascular complex. Am J Physiol Gastrointest Liver Physiol 304(1):G1–G11. https://doi.org/10.1152/ajpgi.00209.2012
15. Richardson DS, Lichtman JW (2015) Clarifying tissue clearing. Cell 162(2):246–257. https://doi.org/10.1016/j.cell.2015.06.067
16. Jing D et al (2019) Tissue clearing and its application to bone and dental tissues. J Dent Res 98(6):621–631. https://doi.org/10.1177/0022034519844510
17. Gomez-Gaviro MV, Sanderson D, Ripoll J, Desco M (2020) Biomedical applications of tissue clearing and three-dimensional imaging in health and disease. iScience 23(8):101432. https://doi.org/10.1016/j.isci.2020.101432
18. Ueda HR et al (2020) Tissue clearing and its applications in neuroscience. Nat Rev Neurosci 21(2):61–79. https://doi.org/10.1038/s41583-019-0250-1
19. Tian T, Yang Z, Li X (2021) Tissue clearing technique: recent progress and biomedical applications. J Anat 238(2):489–507. https://doi.org/10.1111/joa.13309
20. Ariel P (2017) A beginner's guide to tissue clearing. Int J Biochem Cell Biol 84:35–39. https://doi.org/10.1016/j.biocel.2016.12.009
21. Bossolani GDP et al (2019) Comparative analysis reveals Ce3D as optimal clearing method for in toto imaging of the mouse intestine. Neurogastroenterol Motil 31(5):e13560. https://doi.org/10.1111/nmo.13560
22. Li W, Germain RN, Gerner MY (2019) High-dimensional cell-level analysis of tissues with Ce3D multiplex volume imaging. Nat Protoc 14(6):1708–1733. https://doi.org/10.1038/s41596-019-0156-4

Chapter 3

Immunohistochemical Staining Protocol of Enteric Glial Cells in Myenteric Plexi

Aurora Zilli, Silvia Basili Franzin, and Luisa Seguella

Abstract

Despite the development of several genetic approaches to visualize enteric glia within the enteric nervous system (ENS), multiple fluorescence labeling remains an essential tool for simultaneously identifying several target molecules. Fluorescence microscopy is widely used to differentiate between distinct glial subtypes based on their morphological appearance and chemical content, with a high degree of specificity. The dynamic nature of enteric glial cells drives extensive changes in morphology and/or the expression of markers, such as GFAP and S100β, in response to extracellular cues and under distinct physiopathological conditions. These changes depend on factors such as the type and severity of the insult, the time elapsed following the insult, the type of enteric glia, their location within the intestine, and the specific molecular signals they receive. In this context, immunofluorescence provides a combinatorial code that defines the extent of phenotypical changes in distinct glial subtypes based on their protein expression. In this chapter, we describe a protocol designed to immunolabel enteric glia in whole-mount circular muscle myenteric plexus (CMMP) preparations and provide an overview of the main sensitive steps that may affect the quality of the immunolabeling.

Key words Enteric glia, Immunofluorescence, Myenteric plexuses, GFAP, Enteric nervous system

1 Introduction

Enteric glia are a large, heterogeneous population of peripheral neuroglia that encompasses several subpopulations [1, 2]. These are currently defined by morphology and anatomical location within the intestinal wall and are classified into two main types of enteric glia, each with distinct subtypes:

1. *Intraganglionic Glia*: These glia are associated with neuronal cell bodies within the myenteric plexuses (Type I, subtype MP) and submucosal plexuses (Type I, subtype SMP).
2. *Extraganglionic Glia*: These glia reside within the interganglionic fiber tracts connecting myenteric ganglia (Type II), outside the myenteric and submucosal plexuses (Type III, subtype MP

Giuseppe Esposito (ed.), *Enteric Glial Cells: Methods and Protocols*, Methods in Molecular Biology, vol. 2971, https://doi.org/10.1007/978-1-0716-4795-0_3,

and subtype SMP, respectively), in the intestinal mucosa (Type III, subtype mucosa), and associated with nerve fibers in the smooth muscle layers (Type IV).

The advent of genetic tools has enabled a detailed characterization of glial subpopulations, revealing subtype-specific profiles for protein expression [3], transcriptomes, and responsiveness to neuromodulators that differ between intestinal regions and species [1, 4, 5]. However, the extensive plasticity of glial subtypes leads to adaptive phenotypic changes in response to extrinsic cues, including neurotransmitters, neuromodulators, cytokines, bacterial products, and distinct pathophysiological conditions [6, 7]. Capturing these changes provides important information about the ongoing glial plasticity within a more homogeneous population and the degree to which these changes occur in response to extracellular signals.

In this context, immunofluorescence is a valuable immunochemical technique for capturing the variability in the extent to which enteric glia undergo changes in molecular composition and structure in response to physiological stimuli or pathophysiological perturbations of any severity [8]. Most research has focused on enteric glia located within the myenteric plexuses of the colon or distal ileum, as these glia play critical roles in modulating myenteric neuron activity, regulating oxidative stress, providing trophic support, and controlling neuroinflammation, gliogenesis, neurogenesis, and the replenishment of mucosal glia [1].

In this chapter, we describe dissection techniques to achieve whole-mount muscularis preparations of the myenteric plexus and provide an immunohistochemistry protocol to label myenteric glia using the major glial markers (Fig. 1). Each step of the protocol may require modifications according to the desired proteins, which are summarized in the *Notes* section.

2 Materials

Unless specified otherwise, chemicals and reagents are purchased in the highest available quality.

2.1 Sylgard-Coated Petri Dish

- Sylgard 184 Silicone Elastomer kit.
- Petri dish (100 × 15 mm and 35 × 10 mm).
- Plastic jar.
- Plastic pipette.
- Vacuum oven.
- pH meter.

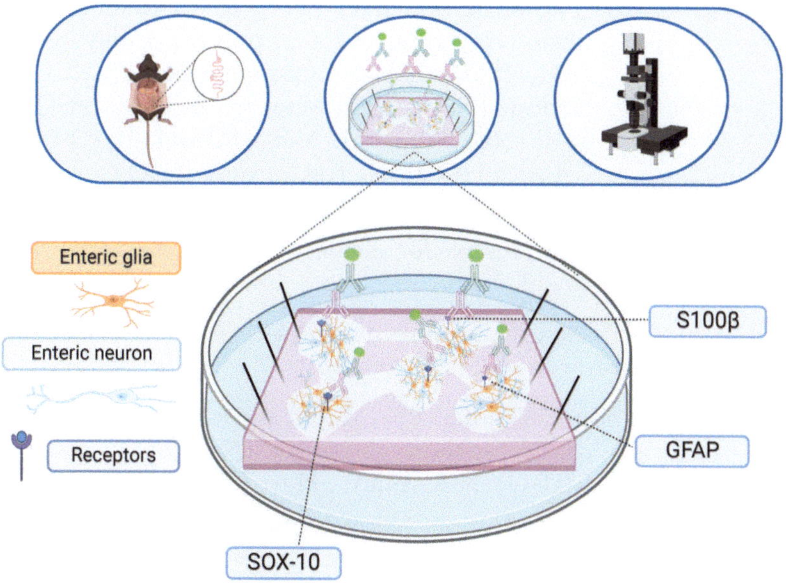

Fig. 1 Schematic illustration of the myenteric plexi immunolabeling procedure

2.2 Mount Circular Muscle Myenteric Plexus (CMMP) Preparations

- Sylgard-coated dish.
- Surgical scissors and forceps.
- Pins 0.20 mm (Fine Science Tools); 0.0125 mm.
- Two curved and straight microdissection forceps (0.18 × 0.2 mm curved tips; 0.1 × 0.06 mm fine 45° curved tips; 0.1 × 0.06 mm fine straight tips).
- Microdissection scissors (0.025 × 0.015 mm tips, 7 mm super fine blades).
- Glass microscope slides (size 25 × 75 mm).
- Microscopy cover glasses.
- DMEM/Ham's F-12 HEPES, no phenol red.
- Dissecting stereo microscope.

2.3 Immunofluorescence

Solutions and Reagents

1× Krebs buffer: Sodium chloride (NaCl) 12.1 mM, potassium chloride (KCl) 0.59 mM, calcium chloride ($CaCl_2$) 0.25 mM, magnesium chloride ($MgCl_2$) 0.12 mM, disodium hydrogen phosphate (Na_2HPO_4) 0.12 mM, sodium bicarbonate ($NaHCO_3$) 21.2 mM, sodium pyruvate ($C_3H_3NaO_3$) 1 mM, glucose ($C_6H_{12}O_6$) 8 mM, HEPES ($C_8H_{18}N_2O_4S$) 10 mM, water for analysis (dH_2O). N.B. Adjust the pH to 7.4 with 10 N NaOH.

4% Paraformaldehyde (PFA)

10× Phosphate buffered saline (PBS): Disodium hydrogen phosphate (Na_2HPO_4) 14.4 g, potassium dihydrogen phosphate (KH_2HPO_4) 2.4 g, potassium chloride (KCl) 2 g, sodium chloride (NaCl) 80 g, water for analysis (dH_2O) 1 L, sodium azide (NaN_3) 2 g. N.B. Mix in 800 mL dH_2O and adjust pH to 7.4 with HCl 10 N. Top up with dH_2O to 1 L.

1× Phosphate buffered saline (PBS): To obtain 1× PBS dilute 10× PBS in water for analysis (dH_2O), ratio 1:10. N.B. Adjust pH to 7.2–7.4.

Phosphate buffered saline with Triton X-100 (T-PBS): 1× PBS, Triton X-100, or Tween 20.

Phosphate buffer (PB): Monobasic sodium phosphate (NaH_2PO_4), dibasic sodium phosphate (Na_2HPO_4), water for analysis (dH_2O). N.B. Adjust pH to 7.2–7.4.

Blocking solution: Normal serum 4%, Triton X-100 0.4%, bovine serum albumin 1%, 1× PBS.

Nucleus staining: DAPI or Hoechst 33342 dye.

ProLong Diamond Antifade mountant.

3 Methods

3.1 Sylgard-Coated Petri Dish

1. Mix 10 parts of Sylgard Base with 1 part of Sylgard Curing Agent.
2. Pour into a Petri dish, filling one-third of the well depth, and set in a level, dust-free location to solidify for 48 h at 20 °C–25 °C (*see* **Note 1**).

3.2 Mount Circular Muscle Myenteric Plexus (CMMP) Preparations

Three different dissection methods may be used to obtain whole-mount circular muscle myenteric plexus (CMMP) preparations [9] (*see* **Note 2**).

3.2.1 Fix Flat Then Peel [10]

1. After euthanasia, intestinal tissue is collected from mice and placed in ice-cold DMEM/Ham's F-12 nutrient mixture or 1× Krebs buffer.
2. The full-thickness tissue is opened along the mesenteric border and pinned flat with mucosa facing up into a Sylgard-coated Petri dish for dissection (100 × 15 mm).
3. The mucosa is removed by cutting at the level of the lamina propria while separating the layers with forceps.
4. Tissues are then flipped into a Sylgard-coated Petri dish for immunohistochemistry (35 × 10 mm) filled with Krebs buffer,

and the longitudinal muscle and serosa are removed by microdissection to obtain live whole-mount preparations with intact myenteric plexus lying atop the preparation.

3.2.2 Fix Flat Then Peel (Without Turning the Tissue) [11]

1. After euthanasia, intestinal tissue is collected from mice and placed in ice-cold DMEM/Ham's F-12 nutrient mixture or 1× Krebs buffer.
2. Tissue is transferred to a Sylgard-coated Petri dish (100 × 15 mm), opened along the mesenteric border, and pinned flat with mucosa facing up. Then, the mucosa, submucosa, and circular muscle were removed with forceps to expose the myenteric plexus.
3. Thus, the tissue is moved into a Sylgard-coated dish for immunohistochemistry (35 × 10 mm) filled with cold Krebs buffer and pinned flat along the edges under light tension.

3.2.3 Peel on Rod [12, 13]

1. After euthanasia, intestinal tissue is collected from mice and placed in ice-cold DMEM/Ham's F-12 nutrient mixture or 1× Krebs buffer.
2. A glass rod is inserted through the lumen of the intestinal segment. The diameter of the glass rod is chosen to fit the diameter of the segment and slightly distend it.
3. Forceps are used to gently perforate the muscularis (but not the mucosa/submucosa) at the mesenteric attachment.
4. A cotton-tipped swab moistened with cold 1× PBS is used to gently peel away the muscular, swabbing circumferentially from the perforated mesenteric edge while the segment is manually held stable on the rod [14].
5. Fully peeled muscular segment is transferred to a Sylgard-bottomed glass dish for immunohistochemistry (35 × 10 mm) filled with cold Krebs buffer and pinned flat along the edges under light tension.

3.3 Immunofluorescence

3.3.1 Sample Fixation

Fixation is an essential step in immunofluorescence staining to preserve the structure and composition of myenteric plexuses (Fig. 2) [15] (*see* **Note 3**).

1. Pin the tissue flat into a Sylgard-coated dish.
2. Fix the tissue with ice-cold 4% PFA for 30 min at 4 °C (*see* **Note 4**).
3. Wash with 1× PBS three times for 10 min at room temperature on the rotator table to remove the fixation buffer (*see* **Note 5**).

3.3.2 Permeabilization

Intracellular antigens become accessible for antibodies by tissue permeabilization, which are otherwise unable to pass through the lipid membranes of the cell (*see* **Note 6**).

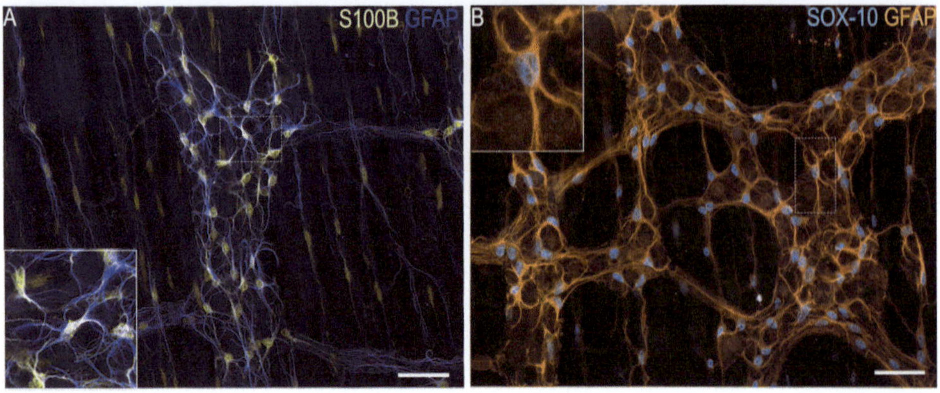

Fig. 2 Immunohistochemical staining of a fixed myenteric plexus. (**a**) The plexus was probed for GFAP expression (blue) and S100β (yellow). (**b**) Expression on enteric glial marker SOX-10 (light blue) and GFAP (orange). Scale bar, 40 μm

1. Rinse the tissue with 0.1% T-PBS (PBS with 0.1% Triton X-100) three times for 10 min on the rotator table at room temperature.

3.3.3 Blocking

To prepare the blocking solution, a protocol previously described by Dr. Brian Gulbransen was used [16]:

1. To make 4% normal serum blocking solution, 600 μL of normal serum and 400 μL of 1× PBS are added to a 15 mL tube.
2. 60 μL of Triton X-100 and 940 μL of PBS are mixed in a 1.5 mL tube and then added to the normal serum solution in a 15 mL tube.
3. 150 mg bovine serum albumin (BSA) is added to the tube, and volume is brought up to 15 mL with 1× PBS.

   The final blocking solution contains 4% normal serum, 0.4% Triton X-100, and 1% BSA in 1× PBS (*see* **Note 7**).
4. Incubate the tissue with the blocking buffer for about 45–60 min on the rotator table at room temperature.

3.3.4 Antibody Incubation (See Note 8)

For indirect immunofluorescence (a fluorochrome-coupled secondary antibody specifically binds to the primary antibody):

1. Determine the optimal antibody dilutions and dilute primary and secondary antibodies in the blocking solution.
2. Incubate the whole-mount preparation of myenteric plexuses with the primary antibody for 2 h at room temperature or overnight at 4 °C on the rotator table (*see* **Note 9**).
3. Wash three times with 1× PBS for 10 min on the rotator table at room temperature.

Table 1
Primary and secondary antibodies used for immunofluorescence labeling

Antigen	Host	Dilution
GFAP	*Chicken*	1:1000
GFAP	*Rabbit*	1: 200
GFAP	*Mouse*	1: 500
SOX-10	*Rabbit*	0.25–2 μg/mL
S100β	*Chicken*	1:500
Anti-Chicken IgG	*Donkey*	1:100–1:800
Anti-Rabbit IgG	*Goat*	1:100–1:800

4. Incubate the whole-mount preparation of myenteric plexuses with secondary antibody for 2 h at room temperature on the rotator table. Incubation must be conducted in the dark to avoid photobleaching. Since fluorophores are photosensitive, all the following steps are carried out in the dark.
5. Wash two times with 1× PBS for 10 min on the rotator table at room temperature.
6. Wash with PB 0.1 M for 10 min on the rotator table at room temperature.

For direct immunofluorescence (a specific primary antibody is linked to a fluorochrome):

1. Determine the optimal antibody dilutions and dilute primary antibodies in the blocking solution.
2. Incubate the enteric plexuses with the primary antibody for 2 h at room temperature on the rotator table.
3. Wash two times with 1× PBS for 10 min on the rotator table at room temperature.
4. Wash with PB 0.1 M for 10 min on the rotator table at room temperature.

   We report in this chapter some of the antibodies used to mark enteric glia (Table 1).

3.3.5 Nucleus Staining and Sample Mounting

To avoid drying out of the samples and guarantee a stable refractive index of the cellular environment (a prerequisite for successful microscopy), the sample needs to be mounted. For this, the sample should be covered with a mounting medium that has a low autofluorescence (*see* **Note 10**).

1. Move the tissues onto a slide, add a few drops of mounting medium, place a coverslip, and seal with nail polish if using an aqueous or glycerol-based mounting medium (*see* **Note 11**).

4 Notes

1. Thoroughly mixed silicone encapsulant may be poured/dispensed directly into the container in which it is to be cured. Care should be taken to minimize air entrapment. A plastic pipette can be used to gently flow air on the silicone encapsulant to remove air bubbles. Dow silicone encapsulants may be either room temperature (25 °C/77 °F) or heat cured in the vacuum oven at 40 °C. Room temperature cure encapsulants may also be heat accelerated for faster cure. Ideal cure conditions for each product are given in the product selection table.
2. The dissection methods should be carefully selected based on the experimental setting and the intestinal region assessed.
3. Fixation may be performed at three distinct stages of the protocol: (1) the intestinal whole-mount preparation is fixed with ice-cold 4% PFA for 30 min before dissection; (2) the tissue is dissected and then fixed with ice-cold 4% PFA for 30 mins; (3) the fixation is performed with ice-cold 4% PFA for 30 min following the dissection and immunolabeling. The fixation method is firmly related to the desired proteins and their location on extracellular or intracellular structures. For extracellular antigens (i.e., receptors), method 3 is suggested. For duodenum and ileum whole-mount preparations, method 2 is suggested.
4. Different fixation methods have different effects on the primary antibody's epitopes (antibody binding sites can be masked or also damaged by fixation). To note, chemical crosslinkers such as formaldehyde preserve cellular morphology by crosslinking proteins via their free amino groups. The antigens are also crosslinked, which might reduce antibody binding. A retrieval pretreatment could be considered. Glutaraldehyde also has a preservative effect on cellular structures, but results in high autofluorescence of the specimen in microscopy. Organic solvents like methanol or acetone have a dehydrogenating effect and precipitate proteins, fixing them in their cellular context. But soluble molecules and many lipid components are lost in the process. To note, fluorescent proteins such as GFP, which are already present in your cells, will be largely destroyed by fixation with organic solvents. If the manufacturer of the primary antibody does not propose a fixative, starting with 4% PFA for 30 min at 4 °C is suitable for various glial antigens. The fixation may be extended until 2 h.
5. Immunostaining quality can be increased by proper washing. 1× PBS is a standard washing buffer. PBS++ contains 1 mM $CaCl_2$ and $MgCl_2$, which is presumed to have a membrane-stabilizing effect, preventing cell detachment. PBS-T, a final

concentration of 0.05–0.5% of the detergent Tween 20 or Triton 100, is added with the aim of increasing the antibody binding specificity.

6. If an organic solvent was used to fix the tissue, cellular membranes became already permeable, and the permeabilization step can be skipped and proceeded directly to blocking. Permeabilization is required with the fixing provided by chemical crosslinkers. Different detergents can be used to permeabilize whole-mount myenteric plexus preparations, including Triton X-100, NP-40, saponin, or Tween 20. Important parameters to set up are detergent concentration and incubation time. To test new antibodies, 0.05–0.1% Triton X-100 in 1× PBS for 15–30 min at room temperature is suggested. To analyze lipid-associated or membrane proteins, the permeabilization step should be carried out carefully or not run at all. A good choice for this purpose is saponin, which selectively removes cholesterol from the plasma membrane, leaving intracellular membranes largely intact. If permeabilization is omitted before antibody staining (only possible by fixation with chemical crosslinkers), extracellular membrane-bound antigens can be specifically marked to distinguish them from the intracellular antigen pool. Nucleic acid dyes such as DAPI or Hoechst are membrane-permeable and do not require permeabilization.

7. It is important that blocking proteins do not originate from the species in which the primary antibody was raised, otherwise the secondary antibody's specificity to the primary antibody will be lost. If you use a secondary antibody, e.g., one produced in a goat against a murine first antibody, an ideal blocking reagent is normal goat serum. Blocking solutions are normally utilized in concentrations around 1% (milk powder, BSA) to 4–5% (normal serum) and diluted in wash buffer. The incubation takes place at room temperature for 30–60 min. Long blocking steps should be avoided, as this can reduce the specific binding of the primary antibody, and therefore reduce the signal. For immunolabeling of myenteric plexus, a blocking solution with 4% normal donkey or goat serum, 0.4% Triton X-100, and 1% BSA in 1× PBS is suggested.

8. Direct or primary immunofluorescence is used mainly to bind target structures in live tissues. The protocol is faster than the indirect version and can provide temporal and spatial analysis. However, it is essential to employ a well-functioning primary antibody with high affinity to its antigen, and validated primary antibodies are expensive. Furthermore, the flexibility in experimental design is reduced in comparison to indirect versions. Indirect or secondary immunofluorescence displays a higher flexibility, and several target structures in the same specimen may be visualized. Moreover, the signal amplification mediated

by the secondary antibody reduces the amount of primary antibody that has to be applied. Otherwise, live analysis and temporal and spatial analysis are not possible.

9. The incubation with the primary antibodies for 48–72 h may be required for some intracellular antigens (i.e., Sox 10, S100β). For extracellular antigens (i.e., receptors), method 3 of fixation with primary antibodies, incubation for 48–72 h, is suggested.
10. Dyes such as Hoechst or DAPI, which enter the nucleus even without permeabilization and intercalate into the DNA, stain the nuclei in 10 min at room temperature. The mounting medium (e.g., Mowiol or Prolong Gold) is used to fix the sample on a microscope slide and prevent its dehydration. Several manufacturers offer mounting media with additives such as DABCO, an anti-fading agent that protects the sample from photobleaching. Depending on the fluorochromes used, some anti-fading agents are more effective than others. The mounting media with added DNA dyes are also available. A hardening mounting medium is allowed to cure overnight, so microscopic examination is possible on the following day. Permanent preparations produced in this way can be practically unlimitedly stored in the dark at room temperature or 4 °C, but remember that the fluorochromes' fluorescence intensity diminishes over time. To mount myenteric plexus preparation, the ProLong Diamond Antifade mountant is suggested.
11. Remove air bubbles trapped between the coverslip and slide.

References

1. Seguella L, Gulbransen BD (2021) Enteric glial biology, intercellular signalling and roles in gastrointestinal disease. Nat Rev Gastroenterol Hepatol 18(8):571–587. https://doi.org/10.1038/s41575-021-00423-7
2. Grubišić V, Gulbransen BD (2017) Enteric glia: the most alimentary of all glia. J Physiol 595(2):557–570. https://doi.org/10.1113/JP271021
3. Boesmans W, Lasrado R, Vanden Berghe P et al (2015) Heterogeneity and phenotypic plasticity of glial cells in the mammalian enteric nervous system. Glia 63(2):229–241. https://doi.org/10.1002/glia.22746. Epub 2014 Aug 26. PMID: 25161129
4. Boesmans W, Cirillo C, Van den Abbeel V et al (2013) Neurotransmitters involved in fast excitatory neurotransmission directly activate enteric glial cells. Neurogastroenterol Motil 25(2):e151-60. https://doi.org/10.1111/nmo.12065. Epub 2013 Jan 2. PMID: 23279281
5. Seguella L, McClain JL, Esposito G et al (2022) Functional Intraregional and Interregional Heterogeneity between Myenteric Glial Cells of the Colon and Duodenum in Mice. J Neurosci 42(46):8694–8708. https://doi.org/10.1523/JNEUROSCI.2379-20.2022. Epub 2022 Nov 1. PMID: 36319118; PMCID: PMC9671584
6. Valès S, Touvron M, Van Landeghem L (2018) Enteric glia: Diversity or plasticity? Brain Res 1693(Pt B):140–145. https://doi.org/10.1016/j.brainres.2018.02.001. Epub 2018 Feb 7. PMID: 29425908
7. Lomax AE, Fernández E, Sharkey KA (2005) Plasticity of the enteric nervous system during intestinal inflammation. Neurogastroenterol Motil 17(1):4–15. https://doi.org/10.1111/j.1365-2982.2004.00607.x
8. Fleming MA 2nd, Ehsan L, Moore SR, Levin DE (2020) The enteric nervous system and its emerging role as a therapeutic target.

Gastroenterol Res Pract 2020:8024171. https://doi.org/10.1155/2020/8024171

9. Ahmadzai MM, Seguella L, Gulbransen BD (2021) Circuit-specific enteric glia regulate intestinal motor neurocircuits. Proc Natl Acad Sci USA 118(40):e2025938118. https://doi.org/10.1073/pnas.2025938118
10. Seguella L, Pesce M, Capuano R et al (2021) High-fat diet impairs duodenal barrier function and elicits glia-dependent changes along the gut-brain axis that are required for anxiogenic and depressive-like behaviors. J Neuroinflammation 18(1):115. https://doi.org/10.1186/s12974-021-02164-5
11. Gulbransen BD, Sharkey KA (2009) Purinergic neuron-to-glia signaling in the enteric nervous system. Gastroenterology 136(4):1349–1358. https://doi.org/10.1053/j.gastro.2008.12.058
12. Gomez-Frittelli J, Hamnett R, Kaltschmidt JA (2024) Comparison of wholemount dissection methods for neuronal subtype marker expression in the mouse myenteric plexus. Neurogastroenterol Motil 36(1):e14693. https://doi.org/10.1111/nmo.14693
13. Smith TH, Ngwainmbi J, Grider JR et al (2013) An in-vitro preparation of isolated enteric neurons and glia from the myenteric plexus of the adult mouse. J Vis Exp 78: 50688. https://doi.org/10.3791/50688
14. Kulkarni S, Micci MA, Leser J et al (2017) Adult enteric nervous system in health is maintained by a dynamic balance between neuronal apoptosis and neurogenesis. Proc Natl Acad Sci USA 114(18):E3709–E3718. https://doi.org/10.1073/pnas.1619406114
15. Ahrends T, Weiner M, Mucida D (2022) Isolation of myenteric and submucosal plexus from mouse gastrointestinal tract and subsequent flow cytometry and immunofluorescence. STAR Protoc 3(1):101157. https://doi.org/10.1016/j.xpro.2022.101157
16. Grubišić V, Gulbransen BD (2017) Enteric glial activity regulates secretomotor function in the mouse colon but does not acutely affect gut permeability. J Physiol 595(11): 3409–3424. https://doi.org/10.1113/JP273492

Chapter 4

Rodent Enteric Glia Culture

Jacques Gonzales, Tony Durand, and Brian D. Gulbransen

Abstract

Enteric glia play fundamental roles in the plasticity and homeostasis of the enteric nervous system. Disruptions in their function contribute to intestinal dysfunction and are associated with gut pathophysiology. To further understand the role of enteric glia in gut physiology, it is essential to use a pure and reliable primary culture method. Here, we describe a protocol for the isolation, purification, enrichment, and culture of enteric glia from mature rodent intestines.

Key words Cell culture, Enteric glia, Glia, Mouse, Rat, Purification, Adult, Isolation

1 Introduction

Enteric glia are close partners of enteric neurons and essential components of the enteric nervous system, which controls digestive functions. Enteric glia maintain intestinal homeostasis through direct interactions with neurons, epithelial cells, and immune cells [1]. Alteration of glial communication contributes to intestinal inflammation, gastrointestinal dysfunctions, visceral pain, and changes in neuroplasticity [1].

Creating simplified and purified culture models of isolated enteric glia is instrumental for understanding the mechanisms of communication between glia and their environment. Various types of enteric glia cultures have been developed, such as immortalized cell lines [2, 3], embryonic cultures of immature enteric glia [4–6], and primary cultures of mature enteric glia [7, 8]. Immortalized cell lines are easy to handle and yield consistent results [2, 3, 9], but they differ significantly from physiological cells [10]. Embryonic enteric glia cultures exhibit high proliferation rates. These cultures maintain the expression of glial markers such as Sox10, GFAP, and S100B while lacking neuronal markers like PGP9.5 and Tubulin b-III [4]. However, embryonic cells differ phenotypically from mature enteric glia. Primary cultures of mature enteric glia best

Giuseppe Esposito (ed.), *Enteric Glial Cells: Methods and Protocols*, Methods in Molecular Biology, vol. 2971,
https://doi.org/10.1007/978-1-0716-4795-0_4,

represent physiological conditions and can be derived from knockout, chemogenetic, or pathological animal models [11]. However, their proliferation is slow and can give rise to fibroblasts after several passages.

Enteric glia in culture express GFAP, S100B, and Sox10 [12–14] and exhibit physiological activities similar to in vivo glia, including calcium activity [13, 15, 16] and the release of proinflammatory cytokines [3, 11], neurotrophic factors [2, 17, 18] and gliomediators such as prostaglandins [11, 19, 20]. However, it is important to note that while these culture models maintain some hallmark features of glia in vivo, they are not entirely physiological and may differ from their in vivo counterparts in many features and functions [2, 21, 22].

Isolating enteric glia in culture is essential for understanding their contribution to gut physiology. In culture, glia can be directly stimulated to assess their activity and mediator release, such as ATP, prostaglandins, or cytokines, under physiological and inflammatory conditions [11, 23]. Enteric glia can be co-cultured with other cell lines to explore the physical and chemical interactions between glia and their surrounding partners. For example, co-cultures of enteric glia with epithelial cell lines enable the study of how enteric glia alter intestinal barrier functions, including epithelium proliferation, barrier healing, and barrier permeability, through the secretion of gliomediators or direct physical interaction with the epithelium [21, 24]. Co-cultures with enteric neurons provide insights into how glia influence neuronal network development and survival [17]. Similarly, co-culturing enteric glia with immune cells allows the investigation of glial contributions to immune responses and intestinal inflammation during diseases [25, 26]. Finally, co-cultures of enteric glia with cancer cells facilitate the study of how enteric glia contribute to cancer cell proliferation and tumor development [5]. These diverse applications make the isolation of enteric glia a valuable tool for understanding their roles in health and disease.

There are several techniques for isolating mature enteric glia [14, 27–29]. This chapter describes an adapted protocol for isolating enteric glia from longitudinal muscle myenteric plexus preparations from adult mice or rats [7, 13, 30]. It includes manual purification, basic maintenance, and characterization of the culture. We refer the user to additional protocols for enteric glia culture from pigs [31] and human surgical resections [7, 32].

2 Materials

2.1 Reagents

1. Poly-D-lysine.
2. Sterilized water.

3. DAPI.
4. Paraformaldehyde 4%.
5. DMEM/F-12.
6. Phosphate buffer saline (PBS).
7. Trypsin-EDTA 0.25%.

2.2 Equipment

1. Fine scissors.
2. Fine curved and straight forceps.
3. 10 cm diameter Petri dishes with Sylgard.
4. Dissection microscope.
5. Brightfield microscope.
6. Culture hood.
7. Sterile pipette tips.
8. Micropipettes.
9. 50 mL conical tube.
10. 15 mL conical tube.
11. 5 mL syringe.
12. GentleMACS™ dissociator.
13. GentleMACS™ C tubes.
14. MACS™ mix tube rotator.
15. Scalpel.
16. Sterile cotton swab.
17. Glass or stainless steel rod of 4 mm diameter for mouse.
18. Glass or stainless steel rod of 20 mm diameter for rat.
19. 24-well culture-coated plates.
20. 12-well culture-coated plates.

2.3 Media and Solutions

1. Digestion solution: 10 mM HEPES, 1× HBSS, 0.13 Wünsch units/mL Liberase TM, 100 Kunits/mL DNAse I (*see* **Note 1**).
2. Blocking solution: PBS, 1% (w/v) bovine serum albumin, 4% (v/v) donkey serum, 1% (v/v) Triton X100; store at 4 °C up to 30 days.
3. Solution of primary antibodies: Blocking solution, 1:1000 mouse anti-alpha smooth muscle actin, 1:1000 chicken anti-GFAP, 1:200 rabbit anti-S100B (*see* **Note 2**).
4. Solution of secondary antibodies: Blocking solution, 1:400 donkey anti-mouse Alexa Fluor 488 nm, 1:400 donkey anti-chicken Alexa Fluor 594 nm, 1:200 donkey anti-rabbit Alexa Fluor 647 nm (*see* **Note 2**).

5. Complete medium: DMEM/F-12, 10% (v/v) heat-inactivated fetal bovine serum (*see* **Note 3**), 50 units/mL penicillin-streptomycin; store at 4 °C up to 30 days (*see* **Note 4**).
6. Mouse medium: DMEM high glucose, 2 mM L-glutamine, 50 units/mL penicillin-streptomycin, 0.2× N2 supplement, 0.2× G5 supplement; store at 4 °C up to 30 days (*see* **Note 4**).
7. Rat medium: DMEM high glucose, 10% (v/v) heat-inactivated fetal bovine serum (*see* **Note 3**), 2 mM L-glutamine, 50 units/mL penicillin-streptomycin; store at 4 °C up to 30 days (*see* **Note 4**).

3 Methods

The methods section describes how to obtain a culture of enteric glia from mice or rats. Most of the steps are similar for both species, albeit with specific media compositions, as described in Subheading 2.3. In addition, the last section presents how the culture can be purified to obtain a purity of 90% or more of enteric glia.

3.1 Preparation for the Culture

All steps in this methods section must be performed in sterile conditions. The preparation steps might be done the day before or the same day as the enteric glia culture. On average, sufficient cell numbers can be obtained from a single animal to produce up to eight wells of enteric glia cultures.

1. Add 200 μL of poly-D-lysine to each well and wait at room temperature (RT) for at least 2 h (*see* **Note 5**).
2. Remove poly-D-lysine and rinse the wells three times with sterile water.
3. Allow the plates to dry under the culture hood for at least 2 h. Allow the coating to dry before introducing any cell or medium culture.
4. Add 500 μL per well of complete medium and store the culture plate in an incubator (37 °C, 5% CO_2) (*see* **Note 6**).

3.2 Tissue Dissection

All steps included in this section must be performed as quickly as possible to ensure cell viability. Reducing the time between euthanasia and the final step of culture is important to obtain a healthy culture.

1. Sacrifice the animal according to the ethical guidelines of the institution (*see* **Note 7**).
2. Take the whole intestine from the animal and keep it in a cup with fresh DMEM/F-12 (*see* **Note 8**).
3. Fill the syringe with DMEM/F-12 and flush out the fecal contents to clean the intestinal tube.

4. Cut the intestine into small pieces of 2–3 cm in length.
5. In a Petri dish filled with fresh DMEM/F-12 medium, insert a piece of colon on the glass/stainless steel rod.
6. Isolate the longitudinal muscle myenteric plexus (LMMP) using the rod dissection technique described by Smith et al. [13] (see publication for a video of the dissection [13]). Briefly, under a dissection scope, remove the mesenteric fat and make a small incision with a scalpel blade on the external layer of the tissue. With a wet cotton swab, gently push the external layer out without disrupting the integrity of the tissue on the rod.
7. Transfer the LMMP preparation to a 2 mL tube containing DMEM/F-12. Store the tube on ice.
8. Re-do **steps 5**–**7** for all sections of the intestine.

3.3 Tissue Dissociation

Tissue dissociation is a crucial step in this protocol. The tissues undergo mechanical and enzymatic dissociation to detach the myenteric ganglia from the tissue structure.

1. Add 2.5 mL of pre-heated digestion solution to a gentle MACS™ C tube.
2. Transfer the dissected LMMP preparations into the gentle MACS™ C tube.
3. Cut the LMMP with fine scissors to mechanically dissociate the tissue.
4. Insert the gentleMACS™ C tube in the gentleMACS™ tissue dissociator and run the pre-set program « m_intestine 01 ».
5. Incubate for 30 min for mouse intestine (37 °C, 5% CO_2), with constant shaking on the gentleMACS™ rotator (4 h for rat intestine).
6. Insert the gentleMACS™ C tube in the gentleMACS™ tissue dissociator and run the program « m_intestine 01 » again.

3.4 Cell Culture and Maintenance

The following step must be done in sterile conditions under a culture hood.

1. Add 10 mL of complete medium to the gentleMACS™ C tube and transfer the cell suspension to a 15 mL conical tube.
2. Centrifuge the solution at 1000 *g* for 5 min (4 °C). Remove the supernatant without disturbing the cellular pellet.
3. Resuspend the cell-containing pellet in 4 mL of complete medium.
4. Add 500 μL of complete medium containing the cells in each coated well (1 mL of final volume in each well).
5. Keep the plate in the incubator (37 °C, 5% CO_2) without disturbing the cells for 72 h.

6. Remove the medium from each well after 72 h and add 1 mL of mouse medium (or rat medium).
7. Refresh half of the medium every 3–4 days by removing 500 μL of medium and adding 500 μL of mouse medium (or rat medium).
8. When the well confluency reaches 80%, treat the cells for the experiment, collect the medium supernatant, or lyse the cells for further molecular study.

3.5 Purification of Glia Culture

This method is an alternative to Subheading 3.4 and increases the glial culture's purity.

1. Add 10 mL of complete medium to the gentleMACS™ C tube and transfer the entire volume to a Petri dish (*see* **Note 9**).
2. Localize ganglia in the solution with a microscope.
3. Pipette the ganglia with a 200 μL micropipette and add them to the culture-coated plate. For an isolated ganglia shape, *see* Fig. 1 and Soret et al. [7].
4. Add 50 ganglia per well in the 24-well plate, and keep the plate in an incubator (37 °C, 5% CO_2).

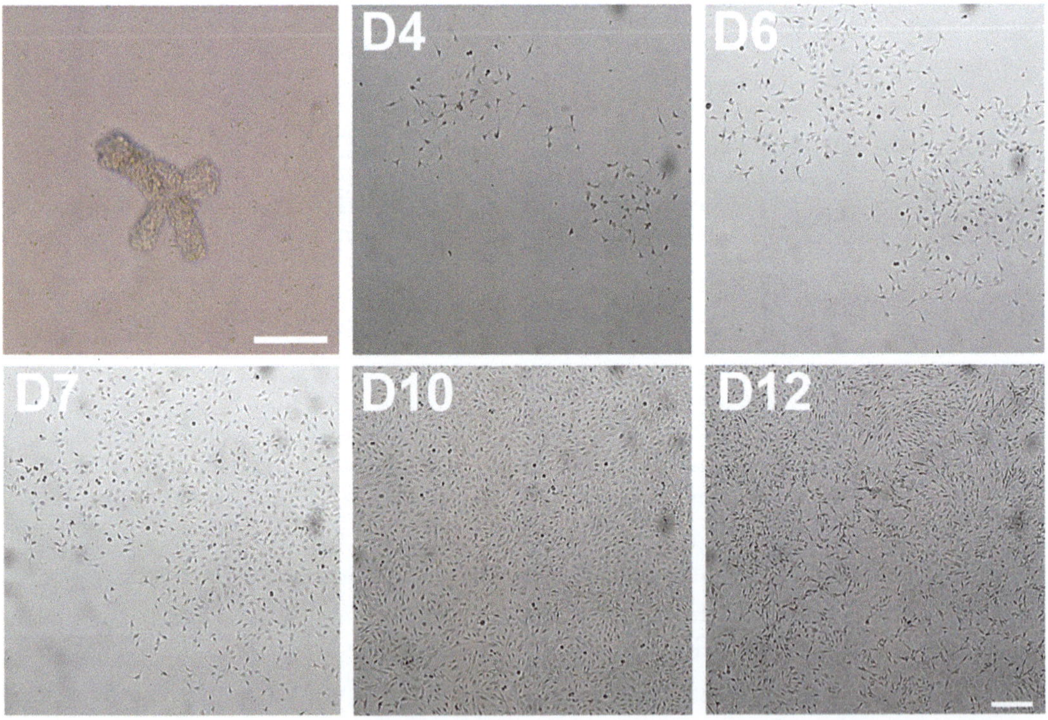

Fig. 1 Culture of mouse enteric glia. Representative mouse ganglia shape for manual pipetting (top left) and development of mouse enteric glia at day 4 (D4), day 6, day 7, day 10, and day 12 post-dissociation observed under brightfield microscopy. Scale 150 μm

5. Gently scratch the fibroblast attached to the well daily to leave only the enteric ganglia. This step will allow users to purify their cultures manually.
6. After 48 h of culture, remove the complete medium and add 1 mL of mouse medium (or rat medium). Continue the following day to scratch the fibroblast.
7. At 80% confluency, perform a passage: Remove the medium and incubate with 100 μL of trypsin-EDTA for 3 min or until the cells are detached (*see* **Note 10**).
8. Add 1 mL of complete medium (containing fetal bovine serum) to each well and centrifuge the cells at 1000 *g* for 5 min at 4 °C.
9. Resuspend the cellular pellet with 16 mL of mouse medium (or rat medium).
10. Add 2 mL per well in a 12-well plate.
11. Re-do **steps 7–10** at 80% confluency to increase cell quantity in larger wells (*see* **Note 11**).

3.6 Verification of Purity

The purity of each glial culture is evaluated in one well following each passage or new culture. For more information and valuable notes, refer to the chapter describing immunohistochemical labeling of enteric glia. The following paragraph briefly describes the protocol for labeling a 24-well plate.

1. Remove culture media and fix with paraformaldehyde 4% for 45 min at room temperature.
2. Wash three times with PBS.
3. Permeabilize the cells with PBS + 0.1% Triton X100 for 5 min.
4. Incubate with blocking solution for 45 min.
5. Incubate the cells with the solution of primary antibodies overnight at 4 °C.
6. On the second day, wash three times with PBS for 10 min.
7. Incubate the cells with the solution of secondary antibodies for 2 h at room temperature.
8. Wash two times with PBS for 10 min.
9. Incubate a solution of DAPI at 1:1000 in PBS for 5 min.
10. Wash two times with PBS for 10 min.
11. Visualize the well under a microscope and assess the percentage of glial cell purity by evaluating the ratio of muscle cells (smooth muscle actin-positive cells) to enteric glia (GFAP and S100B-positive cells) (Fig. 2).

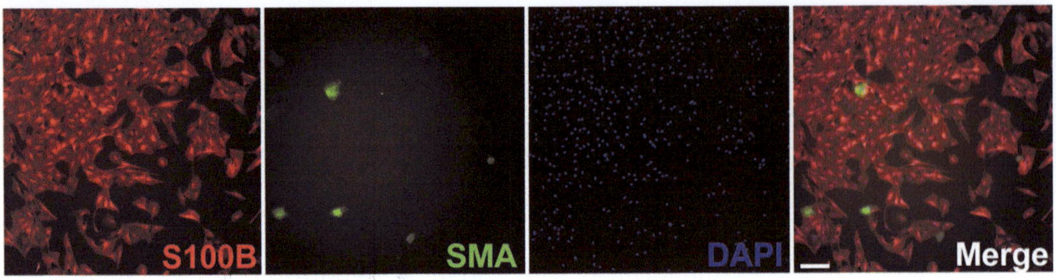

Fig. 2 Representative immunolabeling of enteric glia culture. Characterization of cultures immunolabeled for fibroblasts (smooth muscle actin (SMA), green), enteric glia (S100B, red), and nuclear labeling (DAPI, blue). Scale 150 μm

4 Notes

1. Liberase TH could be used instead of Liberase TM during enzymatic digestion; however, due to its higher activity, the dissociation time should be reduced.
2. The proposed combination of antibodies includes the following antibodies: mouse anti-smooth muscle actin (Abcam, ab7817), chicken anti-GFAP (Abcam, ab4674), rabbit anti-S100B (Abcam, ab52642), donkey anti-mouse Alexa Fluor 488 nm (Jackson ImmunoResearch, 715–545-150), donkey anti-chicken Alexa Fluor 594 nm (Jackson ImmunoResearch, 703–585-155), and donkey anti-rabbit Alexa Fluor 647 nm (Jackson ImmunoResearch, 711–605-152).
3. The serum is heated at 56 °C for 30 min to inactivate the complement system.
4. After 30 days, the serum or supplements contained in the medium will be partially degraded. It is essential not to use an old medium. Check for contamination before using the medium in culture.
5. The coating of poly-D-lysine can be supplemented with laminin (50 μg/mL)—or replaced by 0.5% gelatin in sterile PBS.
6. Preferentially, use the center well of the culture plate to avoid evaporation and fill the unused well with sterile water to keep the plate at a homogenous temperature.
7. Using young animals (4–6 weeks old) increases the density of cells and their proliferation rate; however, the glia culture also works with older animals.
8. This technique can be used for small intestine, colon, or whole intestine glia culture. Culture from the whole intestine allows the user to obtain a higher quantity of enteric glia from one animal; however, using a specific region may be of interest in disease models such as colitis or jejunoileitis.

9. After adding the medium containing the cells to the Petri dish, place it back in the incubator for 30 min (37 °C, 5% CO_2). Over time, the ganglia will settle to the bottom of the Petri dish, making it easier to locate and pipette them.
10. Pure glia cultures from rats could be passed up to passage 4 or 5; however, pure cultures of mouse enteric glia progressively slow their growth after 2–3 passages. Attempting additional passages is not recommended due to the loss of glial marker expression and potential divergence from physiological significance.
11. Pure cultures of mouse enteric glia may stop proliferating even with growth factors in the medium. To prevent this situation, it is advisable to perform a passage every other week.

References

1. Seguella L, Gulbransen BD (2021) Enteric glial biology, intercellular signalling and roles in gastrointestinal disease. Nat Rev Gastroenterol 18:571–587. https://doi.org/10.1038/s41575-021-00423-7
2. Meir M, Kannapin F, Diefenbacher M et al (2021) Intestinal epithelial barrier maturation by enteric glial cells is GDNF-dependent. Int J Mol Sci 22:1887. https://doi.org/10.3390/ijms22041887
3. Costa DVS, Moura-Neto V, Bolick DT et al (2021) S100B inhibition attenuates intestinal damage and diarrhea severity during Clostridioides difficile infection by modulating inflammatory response. Front Cell Infect Microbiol 11:739874. https://doi.org/10.3389/fcimb.2021.739874
4. Landeghem LV, Chevalier J, Mahé MM et al (2011) Enteric glia promote intestinal mucosal healing via activation of focal adhesion kinase and release of proEGF. Am J Physiol Gastrointest Liver Physiol 300:G976–G987. https://doi.org/10.1152/ajpgi.00427.2010
5. Valès S, Bacola G, Biraud M et al (2019) Tumor cells hijack enteric glia to activate colon cancer stem cells and stimulate tumorigenesis. EBioMedicine 49:172–188. https://doi.org/10.1016/j.ebiom.2019.09.045
6. Bondurand N, Natarajan D, Thapar N et al (2003) Neuron and glia generating progenitors of the mammalian enteric nervous system isolated from foetal and postnatal gut cultures. Development 130:6387–6400. https://doi.org/10.1242/dev.00857
7. Soret R, Coquenlorge S, Cossais F et al (2013) Characterization of human, mouse, and rat cultures of enteric glial cells and their effect on intestinal epithelial cells. Neurogastroenterol Motil 25:e755–e764. https://doi.org/10.1111/nmo.12200
8. Teramoto H, Hirashima N, Tanaka M (2022) A simple method for purified primary culture of enteric glial cells from mouse small intestine. Biol Pharm Bull 45:547–551. https://doi.org/10.1248/bpb.b22-00038
9. Meir M, Flemming S, Burkard N et al (2015) Glial cell line-derived neurotrophic factor promotes barrier maturation and wound healing in intestinal epithelial cells in vitro. Am J Physiol-Gastrointest Liver Physiol 309:G613–G624. https://doi.org/10.1152/ajpgi.00357.2014
10. Audesirk GJ (2010) Comprehensive toxicology. Vivo Vitr Syst Neurol Syst:415–432. https://doi.org/10.1016/b978-0-08-046884-6.01325-7
11. Morales-Soto W, Gonzales J, Jackson WF, Gulbransen BD (2023) Enteric glia promote visceral hypersensitivity during inflammation through intercellular signaling with gut nociceptors. Sci Signal 16:eadg1668. https://doi.org/10.1126/scisignal.adg1668
12. Joseph NM, He S, Quintana E et al (2011) Enteric glia are multipotent in culture but primarily form glia in the adult rodent gut. J Clin Invest 121:3398–3411. https://doi.org/10.1172/jci58186
13. Smith TH, Ngwainmbi J, Grider JR et al (2013) An in-vitro preparation of isolated enteric neurons and glia from the myenteric plexus of the adult mouse. J Vis Exp. https://doi.org/10.3791/50688
14. Soret R, Schneider S, Bernas G et al (2020) Glial cell-derived neurotrophic factor induces enteric neurogenesis and improves colon

structure and function in mouse models of Hirschsprung disease. Gastroenterology 159: 1824–1838.e17. https://doi.org/10.1053/j.gastro.2020.07.018
15. Zhang W, Segura BJ, Lin TR et al (2003) Intercellular calcium waves in cultured enteric glia from neonatal Guinea pig. Glia 42:252–262. https://doi.org/10.1002/glia.10215
16. Boesmans W, Martens MA, Weltens N et al (2013) Imaging neuron-glia interactions in the enteric nervous system. Front Cell Neurosci 7:183. https://doi.org/10.3389/fncel.2013.00183
17. Berre-Scoul CL, Chevalier J, Oleynikova E et al (2017) A novel enteric neuron–glia coculture system reveals the role of glia in neuronal development. J Physiol 595:583–598. https://doi.org/10.1113/jp271989
18. Boyen GBTV, Steinkamp M, Reinshagen M et al (2006) Nerve growth factor secretion in cultured enteric glia cells is modulated by Proinflammatory cytokines. J Neuroendocr 18:820–825. https://doi.org/10.1111/j.1365-2826.2006.01478.x
19. Bach-Ngohou K, Mahé MM, Aubert P et al (2010) Enteric glia modulate epithelial cell proliferation and differentiation through 15-deoxy-Δ12,14-prostaglandin J2. J Physiol 588:2533–2544. https://doi.org/10.1113/jphysiol.2010.188409
20. Pochard C, Coquenlorge S, Jaulin J et al (2016) Defects in 15-HETE production and control of epithelial permeability by human enteric glial cells from patients with Crohn's disease. Gastroenterology 150:168–180. https://doi.org/10.1053/j.gastro.2015.09.038
21. Savidge TC, Newman P, Pothoulakis C et al (2007) Enteric glia regulate intestinal barrier function and inflammation via release of S-Nitrosoglutathione. Gastroenterology 132: 1344–1358. https://doi.org/10.1053/j.gastro.2007.01.051
22. Rao M, Rastelli D, Dong L et al (2017) Enteric glia regulate gastrointestinal motility but are not required for maintenance of the epithelium in mice. Gastroenterology 153:1068–1081.e7. https://doi.org/10.1053/j.gastro.2017.07.002
23. Cheadle GA, Costantini TW, Lopez N et al (2013) Enteric glia cells attenuate cytomix-induced intestinal epithelial barrier breakdown. PLoS One 8:e69042. https://doi.org/10.1371/journal.pone.0069042
24. Neunlist M, Aubert P, Bonnaud S et al (2007) Enteric glia inhibit intestinal epithelial cell proliferation partly through a TGF-β1-dependent pathway. Am J Physiol-Gastrointest Liver Physiol 292:G231–G241. https://doi.org/10.1152/ajpgi.00276.2005
25. Kermarrec L, Durand T, Neunlist M et al (2016) Enteric glial cells have specific immunosuppressive properties. J Neuroimmunol 295: 79–83. https://doi.org/10.1016/j.jneuroim.2016.04.011
26. Pabois J, Durand T, Berre CL et al (2020) T cells show preferential adhesion to enteric neural cells in culture and are close to neural cells in the myenteric ganglia of Crohn's patients. J Neuroimmunol 349:577422. https://doi.org/10.1016/j.jneuroim.2020.577422
27. Schonkeren SL, Küthe TT, Idris M et al (2022) The gut brain in a dish: murine primary enteric nervous system cell cultures. Neurogastroenterol Motil 34:e14215. https://doi.org/10.1111/nmo.14215
28. Guyer RA, Stavely R, Robertson K et al (2023) Single-cell multiome sequencing clarifies enteric glial diversity and identifies an intraganglionic population poised for neurogenesis. Cell Rep 42:112194. https://doi.org/10.1016/j.celrep.2023.112194
29. Baghdadi MB, Ayyaz A, Coquenlorge S et al (2022) Enteric glial cell heterogeneity regulates intestinal stem cell niches. Cell Stem Cell 29:86–100.e6. https://doi.org/10.1016/j.stem.2021.10.004
30. Grubišić V, McClain JL, Fried DE et al (2020) Enteric glia modulate macrophage phenotype and visceral sensitivity following inflammation. Cell Rep 32:108100. https://doi.org/10.1016/j.celrep.2020.108100
31. Caldwell ML, Cook CA, Mariant CL et al (2024) Protocol to culture enteric glial cells from the submucosal and myenteric plexi of neonatal and juvenile pig colons. STAR Protoc 5:103057. https://doi.org/10.1016/j.xpro.2024.103057
32. Cirillo C, Sarnelli G, Turco F et al (2011) Proinflammatory stimuli activates human-derived enteroglial cells and induces autocrine nitric oxide production. Neurogastroenterol Motil 23:e372–e382. https://doi.org/10.1111/j.1365-2982.2011.01748.x

Chapter 5

Isolation of Distinct Populations of Enteric Glia from the Muscularis Externa and Mucosa of the Mouse Intestine

Anoohya N. Muppirala and Meenakshi Rao

Abstract

Enteric glia, integral to the enteric nervous system (ENS), are crucial for intestinal motility, secretomotor function, and immunity. These glia occupy diverse niches from the serosa to the lumen, yet their transcriptional diversity across these compartments remains incompletely understood. Traditionally, studies on enteric glia have mostly focused on the myenteric plexus, omitting the large populations of glia located in the circular muscle and mucosa. Here, we present a method to rapidly isolate enteric glia from the muscularis externa and mucosal compartments of the mouse intestine in tandem, allowing direct comparative and compartment-specific analyses of each population. This protocol circumvents the need for traditional and labor-intensive longitudinal muscle myenteric plexus (LMMP) peeled tissue preparations by enabling rapid cell isolation suitable for many downstream applications, ranging from transcriptional profiling to cell culture. We detail a protocol for mechanically separating adult small intestinal layers, dissociating cells from the mucosal and muscularis compartments, and sorting these cells by flow cytometry. This method is applicable to both small and large intestines and has been validated in mice from weaning to adulthood. The ability to isolate distinct populations of enteric glia will enable functional interrogation of their niche-specific roles, advancing our understanding of their diverse contributions to ENS biology and pathology.

Key words Enteric glia, Glial cell isolation protocol, Intramuscular glia, Mucosal glia, Muscularis glia, Enteric nervous system

1 Introduction

Enteric glia closely associate with neurons of the enteric nervous system (ENS), both with their somas within ganglia and with their extensive projections outside ganglia that course throughout the muscularis, submucosal and mucosal layers of the digestive tract (Fig. 1). These glia are essential for ENS function and play important roles in intestinal motility, secretomotor function, and immunity [1, 2]. To date, most studies of enteric glia have either examined them as a whole or focused on those present in

Giuseppe Esposito (ed.), *Enteric Glial Cells: Methods and Protocols*, Methods in Molecular Biology, vol. 2971,
https://doi.org/10.1007/978-1-0716-4795-0_5,

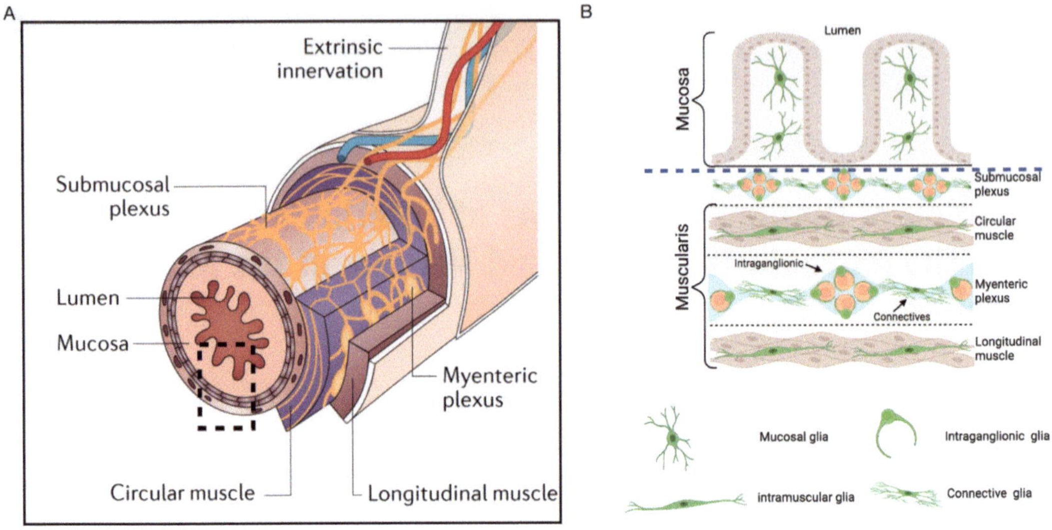

Fig. 1 Enteric glia are morphologically diverse and located in different compartments across the laminar structure of the intestine. (**a**) Schematic illustrating the organization of the enteric nervous system within the mouse intestine (adapted from [7]). (**b**) Schematic of cross-section of mouse small intestine demonstrating morphological diversity of glia along the radial axis in distinct compartments from the outer muscularis externa to the mucosa. The dashed purple line indicates the point of mechanical separation between the two layers of the intestine in this protocol. Schematic created with Biorender

"LMMP" preparations commonly used in the field, which contain the longitudinal muscle with adherent myenteric plexus. LMMP preparations do not include the large populations of enteric glia present in the mucosa, submucosal plexus, or circular muscle, leaving the biology of many of these cells understudied.

Glia in the ENS also exhibit transcriptional, morphological and phenotypic heterogeneity that varies by regional identity along the gastrointestinal tract as well as along the radial axis from serosa to lumen. For example, upon systemic lipopolysaccharide (LPS) administration to rats, markers of reactive glia are upregulated within the myenteric plexus but not within the mucosa or submucosa [3]. Conversely, in germ-free mice raised in the absence of microbes, mucosal glia, but not myenteric plexus glia, are mispositioned [4]. Thus, interrogating the biology of enteric glia in a comparative and compartment-specific manner is likely to be informative. Here, we present a method developed to isolate murine enteric glia in tandem from the muscularis externa and mucosal compartments. In contrast to LMMP preparations, which require painstaking and precise microdissections of tissue layers [5, 6], this method is optimized for rapid cell isolation and downstream applications that require live cells. The resulting cell suspensions can be utilized for transcriptional profiling, in vitro cell culture, or functional interrogation. The method below is readily adaptable to the small and large intestines and has been validated in mice from weaning to adulthood, making it widely applicable. Here we describe the necessary steps in detail including solution preparation (Materials 2), mechanical separation of adult small intestinal tissues (Subheading 3.1), and the subsequent disassociation protocols for the mucosal and muscularis compartments, in Subheading 3.2 and Subheading 3.3, respectively. For studies directly comparing enteric glia from the two compartments, these dissociations should be done in tandem with the steps staggered to maintain similar sorting conditions and minimize variability.

2 Materials

1. 10× Hanks' buffered saline solution (HBSS).
2. 15- and 50 mL polypropylene tubes.
3. 100× Penicillin–Streptomycin (Pen/Strep).
4. 37 °C water bath with capacity to incubate 15 mL tubes.
5. Centrifuge with cooling capability and rotor with capacity for 15 mL tubes.
6. Heat-inactivated fetal bovine serum (FBS).
7. Dithiothreitol (DTT).
8. 0.5 M ethylenediaminetetraacetic acid (EDTA, pH 8).
9. 10× Phosphate-buffered saline.
10. Sterile deionized water.
11. Crochet needle (1.6-mm hook for adult mice—can adjust depending on size of intestinal opening).
12. Small cotton swabs: 15 cm wood handle, small tip—autoclaved.
13. Razor blades.
14. Large rectangular ice bucket (for dissection).
15. Small ice bucket (for transfer to and from centrifuge).
16. 100- and 70 μm cell strainers fitted for 50 mL Falcon tubes.
17. Live-dead stain: Propidium iodide.
18. Digestion medium: HEPES buffered saline 2×, 25 mM $CaCl_2$, Collagenase A, Dispase II, DNAse I, deionized water.
19. Trypsin-EDTA 0.25% phenol red.
20. 10 cm Petri dishes.
21. 24-well non-TC-treated polystyrene plates.
22. Intestinal flushing apparatus: 10 mL syringe, tubing detached from butterfly needle, and unfiltered 200 μL pipette tips.
23. Sterile fluorescence-activated cell sorting (FACS) polystyrene tubes.

24. Individually wrapped 5 and 10 mL stripette polystyrene serological pipettes.
25. 1000 μL pipette and filtered tips.
26. 200 μL pipette and filtered tips.
27. Fine tweezers and dissecting scissors.
28. 70% Ethanol (EtOH).
29. Disposable polyethylene transfer bulb pipettes.
30. DNAse 1.
31. Tris(2-carboxyethyl)phosphine [TCEP].
32. RNeasy Micro Kit.
33. RNeasy lysis buffer (RLT).
34. RNeasy buffer RPE (RPE).
35. RNeasy buffer RW1 (RW1).
36. Digestion medium (100 mL): 50 mL of HEPES 2×, 5 mL of 25 mM $CaCl_2$, 100 mg Collagenase A, 290 mg Dispase II, 5 mg DNAse 1, 45 mL of sterile water (*see* **Note 1**). 100 mL of digestion medium frozen into aliquots of 10 mL each at −20°C. Pre-warm a 10 mL aliquot of digestion medium in a 37°C water bath for 5–10 min prior to beginning the tissue dissection. Split into 5 mL aliquots pre-labeled with "muscularis" and "mucosa."
37. 1× HBSS from 10× stock: 20 mL of 10× stock to 180 mL of deionized water.
38. 100 mL Pen/Strep-HBSS solution: 1000 μL of 100× Pen/-Strep to 99 mL of 1× HBSS.
39. Set the centrifuge to cool down to 4°C for later spin down steps.
40. 50 mL of FACS buffer: 1 mL of FBS to 1× PBS and 1 mM EDTA.
41. Chilled 0.01 M DTT: 100 μL of DTT to 9.9 mL of 1× HBSS and keep on ice.
42. Chilled 5 mM EDTA in PBS: 44.5 mL of dd H_2O + 5 mL of 10× PBS + 0.5 mL of 0.5 M EDTA.
43. Pen/Strep-HBSS solution (split into two 50 mL conical tubes and a 10 cm Petri dish containing ~20 mL), DTT, crochet needle, razor blade, and syringe for intestinal flushing (filled with 1× Pen/Strep-HBSS solution) and three 10 cm Petri dishes on ice (*see* **Note 2**).
44. Flushing syringe (cut butterfly needle off, thus exposing the tubing [~6 cm]). Then, attach the tubing to the 10 mL syringe Luer Lock. Push a non-filter 200 μL pipette tip onto the

exposed end of the tubing to create a seal. Fill the flushing apparatus with chilled HBSS +1× Pen Strep and submerge in an ice bucket until use.

3 Methods

3.1 Mechanical Separation of Intestinal Tissue

1. Euthanize the mouse by cervical dislocation, place it in a supine position on ice (on its back), spray liberally with 70% EtOH, and carefully make a vertical incision along the midline across the entire abdominal wall through to the peritoneal cavity.
2. Gently move or lift the liver to expose the stomach. Cut the stomach in half, holding on to the glandular stomach segment attached to the small intestine.
3. Carefully pull up while simultaneously trimming away fat and the mesentery attached to the intestinal wall (*see* **Note 3**) until the ileocecal junction is visible. Avoid force as this can damage the *muscularis externa*.
4. Cut the small intestine proximal to the ileocecal junction and place the now isolated stomach to the small intestinal portion of the digestive tract in the unlabeled Petri dish containing chilled Pen/Strep-HBSS.
5. Isolate ~10 cm of the distal small intestine (measuring backward from the ileocecal junction), carefully trim off the remaining mesentery, and discard the remaining segment of proximal small intestine that is still attached to the stomach.
6. Gently insert the flushing syringe pipette tip into the distal ileum and flush the intestine until clear of all debris and fecal content.
7. Taking the crochet needle, gently insert it into the lumen of the distal intestine segment and slide on until the crochet needle is ¾ covered. Keep the overhang of the intestine submerged in ice-cold Pen/Strep-HBSS (Fig. 2a).
8. If right-handed, hold the tip of the intestinal segment that is furthest from the end of the crochet needle (beginning of the segment that was placed onto the needle) with your left thumb.
9. Holding the razor blade parallel to the length of the intestine, lightly make an incision at the intestinal end adjacent to your left thumb, only through the muscle layer. If the crochet needle becomes exposed by this incision, create a new incision as this indicates the cut was too deep (*see* **Note 4**) (Fig. 2b).
10. Take a small dry cotton swab and gently peel the muscle at the incision with an upward movement, perpendicular to the crochet needle, while simultaneously using your left thumb to rotate the needle and intestine downward in the opposite direction (Fig. 2c).

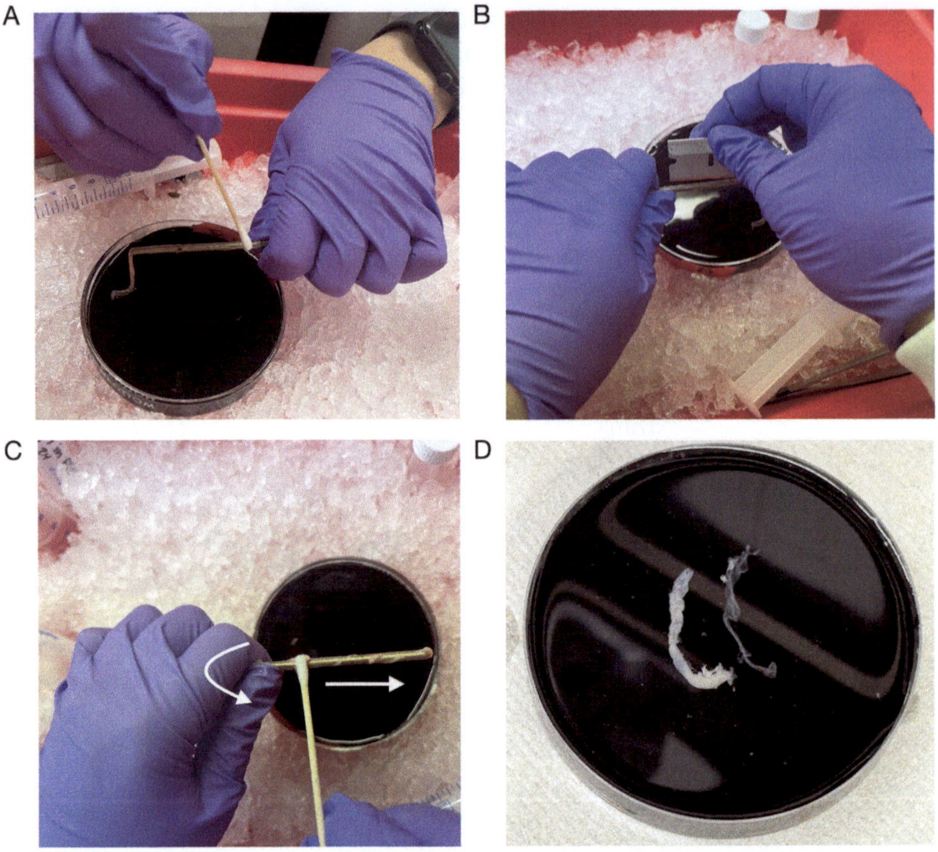

Fig. 2 Mechanical separation of intestinal tissue for isolation of glia from muscularis and mucosal compartments. (**a**) Positioning of the distal ~10 cm of small intestine on the crochet needle with the thumb at the anticipated incision site. (**b**) Position of the blade parallel to the intestinal tissue on the crochet needle. (**c**) Circular motion of pulling the intestine in a counterclockwise fashion while simultaneously peeling back the muscle with a cotton swab. (**d**) Example of final muscle strip (right) next to mucosa (left). The technique allows for reliable separation of compartments as tubes from starting segments up to ~10 cm long. Tissue in these representative images was obtained from an 8-week-old male C57Bl6/J mouse. With practice and proficiency, the entire procedure from euthanasia of the animal to isolation of the two tissue compartments, as shown in (**d**), requires less than 3 min

11. When the crochet needle has been rotated 360° and the muscle wall is detached from the mucosa, place your left thumb on the exposed mucosa and gently pull more of the overhanging intact intestine onto the crochet needle. Repeat **steps 9** and **10**, rotating and pulling the tissue until the entire muscularis layer has been peeled off, leaving the mucosa on the crochet needle.
12. Immediately submerge the peeled muscularis segment and the residual mucosal segment on the crochet needle into their respective Petri dishes with Pen/Strep-HBSS and move on to the dissociation steps (Fig. 2d).

3.2 Dissociation of Mucosa

1. Cut the mucosa along the longitudinal axis to expose the epithelium and place the tissue in a chilled tube containing 0.01 M DTT. Shake vigorously for 30 s and then allow the tissue to sit submerged in solution on ice for 5 min.
2. Transfer the mucosal tissue with tweezers to 5 mM EDTA in a 15 mL conical tube and let sit for another 5 min on ice.
3. Shake vigorously for 30 s.
4. Transfer mucosal tissue to chilled Pen/Strep-HBSS solution to rinse. Repeat once more to thoroughly wash off EDTA and DTT.
5. Decant the wash solution from the Petri dish containing mucosal tissue and replace it with a small volume of the preheated digestion media.
6. Cut mucosa into 1–2 mm fragments. It is crucial that the tissue segments are sufficiently small. Then, using the transfer bulb pipette, aspirate the tissue segments along with the digestion medium and transfer them into the conical tube containing the remainder of the 5 mL prewarmed digestion medium. Incubate in a 37 °C water bath for 30 min, gently inverting every 5 min. Triturate with a 5 mL serological pipette midway through this incubation.
7. Following the 30 min digestion, triturate again with a 5 mL serological pipette. Spin down at 500 g for 5 min at 4 °C.
8. Resuspend in 5 mL of 0.25% Trypsin-EDTA in HBSS (Ca^{2+}/Mg^{2+} free) and incubate at 37 °C for 5 min. Add 10 mL of FBS to stop the reaction.
9. Spin cells down at 500 g for 5 min at 4 °C.
10. Decant supernatant and resuspend the cell pellet in 0.1 mg/mL of DNAse I in HBSS (~2 mL) and incubate at 37 °C for 10 min (*see* **Note 5**).
11. Spin down cells at 4 °C.
12. Resuspend the cell pellet in 5 mL of chilled FACS buffer, triturate thoroughly with a 1000 μL pipette tip, and then filter the cell suspension through a 100 μm cell strainer into a 50 mL conical tube chilling on ice. Repeatedly tap the 50 mL conical tube in the ice bucket while forcefully dispensing 10 mL of FACS buffer through the strainer to dislodge and collect cells caught in the strainer. Do this 1 mL at a time with a 1000 μL pipette.
13. Spin cells down at 500 g for 5 min at 4 °C. Decant supernatant. *Optional:* Perform **steps 14–19** for cell sorting and RNA extraction.
14. Resuspend in 1 mL of FACS buffer (*see* **Note 6**) and transfer to a FACS tube.

15. Add propidium iodide (PI) to a final concentration of 1 μg/mL (1:1000) before sorting.
16. Mix the cell suspension well with a 1000 μL pipette prior to sorting and ensure that the sample is continuously agitated during sorting.
17. Sort cells into 350 μL of buffer RLT + 3.5 μL of TCEP per well in a 24-well plate.
18. Transfer sorted cells into a 2 mL microcentrifuge tube, vortex, and immediately move on to the downstream application. For gene expression studies, extract and clean up RNA using the Qiagen RNAeasy Micro kit per the manufacturer's guidelines (*see* **Note 7**).
19. RNA can be stored at −80 °C until ready to be used.

3.3 Dissociation of the Muscularis Tissue

(To be done in parallel with Subheading 3.2 if collecting glia from both compartments)

1. Gently lift the muscularis tissue from the Petri dish, decant Pen/Strep-HBSS, and add a small volume (~1 mL) of pre-warmed digestion media to the dish, positioning the Petri dish at an angle to collect digestion media in one corner.
2. Cut the muscularis strip into small pieces (1–2 mm fragments) in the medium. It is crucial that the tissue segments are sufficiently small. Then, using the transfer bulb pipette, aspirate the tissue segments along with the digestion medium and transfer them into the conical tube containing the remainder of the 5 mL of pre-warmed digestion medium. Incubate for 30 min at 37 °C, gently inverting every 5 min. Triturate with a 5 mL serological pipette halfway through incubation.
3. After digestion has finished, triturate with a 5 mL serological pipette. Spin down at 500 g for 5 min at 4 °C (*see* **Note 8**).
4. Decant supernatant and resuspend the cell pellet in 5 mL of 0.25% Trypsin-EDTA in HBSS (Ca/Mg free) and incubate at 37 °C for 5 min. Add 10 mL of FBS to stop the reaction.
5. Spin cells down at 500 g for 5 min at 4 °C.
6. Resuspend the cell pellet in 5 mL of chilled FACS buffer, triturate well with a 1000 μL pipette tip, and then filter the cell suspension through a 100 μm cell strainer into a 50 mL conical tube chilling on ice. Repeatedly tap the 50 mL conical tube in the ice bucket while forcefully dispensing 10 mL of FACS buffer through the strainer to dislodge and collect cells caught in the strainer. Do this 1 mL at a time with a 1000 μL pipette.
7. Spin cells down at 500 g for 5 min at 4 °C. Decant supernatant.

8. Resuspend in 250 μL of FACS buffer and transfer to a FACS tube.
9. Add propidium iodide (PI) to a final concentration of 1 μg/mL (1:1000) before sorting.
10. Mix the cell suspension well with a 1000 μL pipette prior to sorting and ensure that the sample is continuously agitated during sorting.
11. Sort cells into 350 μL of Buffer RLT + 3.5 μL of TCEP per well in a 24-well plate (*see* **Notes 9** and **10**).

4 Notes

1. Digestion media can be made and stored frozen in advance of the experiment (stable for several months at −20 °C). 5 mL of digestion media should be used per compartment of stripped tissue for 1–5 adult mice, so aliquots can be made based on the experiment size desired by the user. For example, if mucosa and muscle from only the ileum of one mouse will be taken, then one 10 mL aliquot is sufficient. If the mucosa and muscle of six mice are pooled, then 20 mL will be necessary—10 mL for the pooled mucosa and 10 mL for the pooled muscularis. For every five tissue segments (i.e., five mice), 5 mL of digestion media is required.
2. For preparing all reagents in a large ice bucket, it is helpful to label Petri dishes with the intestinal segment of interest. If stripping both mucosa and muscle, leave one Petri dish unlabeled for initial intestinal dissection, then label one Petri dish with “muscularis” and one with “mucosa,” and place all three plates on ice and fill with Pen/Strep-HBSS solution. This organizational scheme will minimize errors while moving forward during dissociation steps. Place the DTT and EDTA solutions on the side (in ice) with the mucosal Petri dish, as they will not be used for the muscularis.
3. While removing the small intestine, place the exposed tissue into the unlabeled Petri dish. When the intestine is pulled more than 5 cm from where the scissors are cutting the mesentery, adjust the hold of the tweezers to pull as close to the cutting site as possible. This minimizes the tearing of the muscle wall.
4. To make a clear incision, gently stroke the razor blade against the intestinal muscularis parallel to the length of the intestine and the crochet needle. Then, take the razor blade to gently push the top incision fold of the tissue perpendicularly to the intestine to expose the mucosa underneath. The amount of pressure to apply differs between the small and large intestines. The small intestine requires less force as the muscle wall is

thinner. Also, ensure that the intestinal tissue that is not on the crochet needle remains submerged in the Petri dish in cold Pen/Strep-HBSS solution. Additionally, the cotton swab size can be optimized by the user. Generally, a large cotton swab is ideal for the upper small intestine after incision, as the mucosa often generates moisture, posing a challenge to peeling back the muscle. A large cotton swab may also be ideal for the distal colon since the muscle wall is thicker and requires more friction for separation. Generally, a small cotton swab is the most recommended for the initial peeling of the muscle after the incision to avoid tearing the mucosa.

5. The DNAse incubation step is not always necessary. In fact, if the cells do not appear to clump, this step should be eliminated to increase the viability and health of the cells. This step can be helpful for dissociating multiple mucosal segments with large cell pellets.
6. The final suspension for the mucosal cells in the FACS tube is very dependent on cell yield through the dissociation, which is also affected by the number of pooled samples. Initially, resuspension in 250 μL of FACS buffer + PI is sufficient. If the cell solution is opaque, then more volume may be added. The suspension should be semitransparent for optimal sorting. The mucosa typically requires 500–1000 μL of FACS buffer for final suspension, while the muscularis requires 250–500 μL.
7. The RNAEasy Micro kit is optimal for low cell yields (<500,000 cells), which is the range for glia isolated from one to five pooled samples of muscularis or mucosa. It is also ideal to perform this step immediately post-sorting to eliminate one freeze–thaw step, which will increase RNA yield. Alternatively, cells can be sorted directly into a plate with DMEM culture media for single-cell sequencing applications.
8. If performing the muscularis and mucosal dissociations simultaneously, the muscularis strip can be placed on ice until the mucosal digestion is completed, before spinning down cells. This will align the dissociation steps.
9. For sorting mucosal samples, a FACS machine with a 130 μm nozzle is ideal to prevent clogging. The SONY SH800S is suitable and was used successfully with this protocol for isolating enteric glia for transcriptional profiling.
10. For sorting muscularis glial cells from weanling to adult mice, the expected percentage of cells is ~3–4% of the total, with an average of 10,000 glial cells within 15 min of sorting. Total cell yield will vary depending on the size and age of the animals used, with standard adult mice from 8 to 12 weeks yielding the most cells. For sorting mucosal glial cells, the expected percentage is ~0.05–1% of total cells, with 10,000 glial cells taking on average 30 min for sorting.

References

1. Rosenberg HJ, Rao M (2021) Enteric glia in homeostasis and disease: from fundamental biology to human pathology. iScience 24(8): 102863. https://doi.org/10.1016/j.isci.2021.102863
2. Seguella L, Gulbransen BD (2021) Enteric glial biology, intercellular signalling and roles in gastrointestinal disease. Nat Rev Gastroenterol Hepatol 18(8):571–587. https://doi.org/10.1038/s41575-021-00423-7
3. Rosenbaum C, Schick MA, Wollborn J et al (2016) Activation of Myenteric Glia during Acute Inflammation In Vitro and In Vivo. PLoS One 11(3):e0151335. https://doi.org/10.1371/journal.pone.0151335
4. Kabouridis PS, Lasrado R, McCallum S et al (2015) Microbiota controls the homeostasis of glial cells in the gut lamina propria. Neuron 85(2):289–295. https://doi.org/10.1016/j.neuron.2014.12.037
5. Bannerman PG, Mirsky R, Jessen KR (1987) Analysis of enteric neurons, glia and their interactions using explant cultures of the myenteric plexus. Dev Neurosci 9(4):201–227. https://doi.org/10.1159/000111623
6. Smith TH, Ngwainmbi J, Grider JR et al (2013) An in-vitro preparation of isolated enteric neurons and glia from the myenteric plexus of the adult mouse. J Vis Exp 78:10.3791/50688
7. Rao M, Gershon MD (2016) The bowel and beyond: the enteric nervous system in neurological disorders. Nat Rev Gastroenterol Hepatol 13(9):517–528. https://doi.org/10.1038/nrgastro.2016.107

Chapter 6

Isolation and Dissection Procedures for EGC Isolation by Human Mucosal Biopsies

Giovanni Sarnelli, Irene Palenca, Maria Sole Facioni, Pierluigi De Pierro, and Marcella Pesce

Abstract

This chapter provides a detailed guide for isolating and dissecting enteric glial cells (EGCs) from human mucosal biopsies. EGCs play crucial roles in gut homeostasis and neuroimmune communication, making their isolation a valuable endeavor in neurogastroenterology research. The method outlined here aims to enable competent scientists, even those unfamiliar with the technique, to successfully isolate EGCs at their first attempt. It outlines a systematic approach for the collection, digestion, mechanical dissection, and purification of cells from human mucosal biopsies, primarily focused on gastrointestinal tissue.

During the first step, tissue collection, the collaboration with clinical teams ensures timely acquisition of biopsies with minimal trauma. Biopsies are then promptly transferred to a cold solution to maintain viability during transport. The subsequent enzymatic digestion involves immersion of biopsies in a solution containing collagenase and Dispase II, optimizing conditions for maximal cell yield. Gentle agitation aids enzyme penetration without causing cellular stress. Mechanical dissection follows, where tissues are minced into uniform fragments using sterile instruments, maintaining sterile conditions to prevent contamination. Gentle trituration further disperses cells without introducing air bubbles. Finally, cell purification involves filtration to remove debris, followed by centrifugation to pellet cells. Careful resuspension in an appropriate medium ensures viability for downstream applications. Overall, this method offers a meticulous approach to isolate viable cells from human mucosal biopsies for further analysis.

Key words Enteric glial cells, Isolation, Dissection, Mucosal biopsies, Enzymatic digestion, Cell purification, Tissue processing, Cell characterization

1 Introduction

Neurogastroenterology stands at the intersection of neuroscience and gastroenterology, exploring the intricate connections between the nervous system and the gastrointestinal tract.

The myenteric and submucosal plexuses are distinguished by their ganglia, which harbor enteric neurons and enteric glial cells (EGCs) in a ratio of approximately 1:7 [1]. Within the enteric nervous system (ENS), these ganglia play pivotal roles in regulating

Giuseppe Esposito (ed.), *Enteric Glial Cells: Methods and Protocols*, Methods in Molecular Biology, vol. 2971, https://doi.org/10.1007/978-1-0716-4795-0_6,

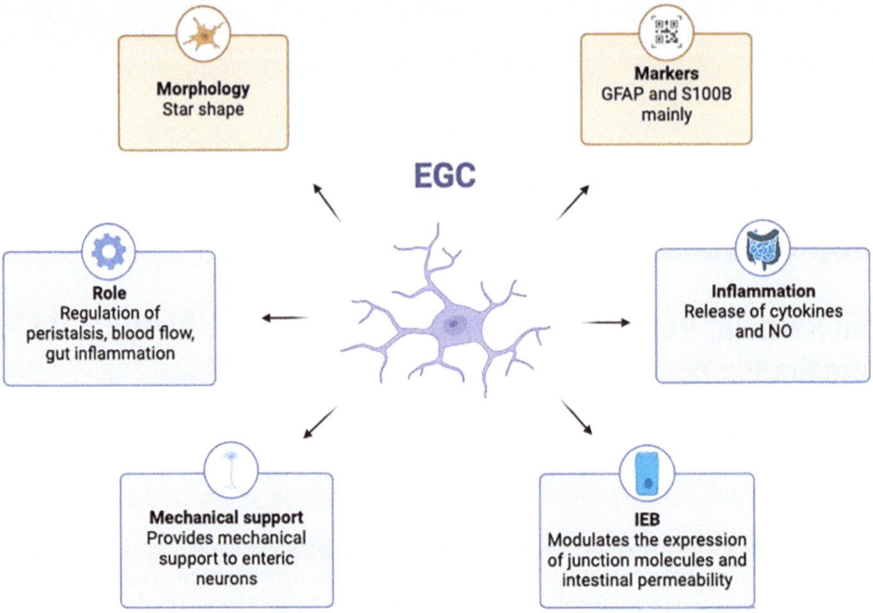

Fig. 1 Schematic view of an EGC

fundamental gut functions such as peristalsis, secretion, blood flow, and immune/inflammatory processes modulation.

EGCs exhibit characteristic morphology resembling stars (*see* Fig. 1) and contain intracellular arrays composed of 10-nm filaments primarily constituted of glial fibrillary acidic protein (GFAP) [2]. Originally identified through methylene blue staining on full-thickness preparations, EGCs are now commonly recognized using specific markers like the S100B protein and GFAP [2].

Apart from providing mechanical support to enteric neurons, EGCs release diverse factors that influence the development, survival, and differentiation of peripheral neurons. Recent insights have expanded our understanding of EGCs beyond mere structural support, highlighting their intricate involvement in maintaining intestinal homeostasis [3, 4] (Table 1).

Studies have demonstrated that EGCs regulate intestinal epithelial barrier (IEB) functions, with animal models showing enhanced intestinal vascular permeability and increased IEB paracellular permeability upon ablation of the enteroglial network [5, 6]. Additionally, in vitro experiments have elucidated that EGCs contribute to reducing IEB permeability through the release of S-nitrosoglutathione and the modulation of zonulin-1 and occludin expression [4].

Furthermore, EGCs respond to inflammatory insults by serving as antigen-presenting cells and promoting a diverse release of cytokines in the gut milieu [7]. They express interleukin-6 (IL-6)

Table 1
Introduction to EGCs

Element	Description
Morphology	Stellate shape, containing 10-nm intracellular filaments mainly composed of GFAP
Markers	S100B protein and GFAP
Mechanical support	Provide mechanical support to enteric neurons
EGCs' role	Regulation of fundamental gut functions such as peristalsis, secretion, blood flow, and immune/inflammatory processes
Intestinal Epithelial Barrier (IEB)	Regulate IEB functions by reducing permeability through the release of S-nitrosoglutathione and modulating the expression of zonulin-1 and occludin
Response to inflammatory insults	Function as antigen-presenting cells and promote the release of cytokines such as IL-6 and IL-1b, in addition to the release of NO

and IL-1b and possess the machinery for the time-delayed and micromolar release of nitric oxide (NO), a critical signaling molecule involved in host immune defense and inflammatory responses against pathogens [7, 8].

Central to advancing our understanding in this field is the ability to isolate and dissect EGCs from human mucosal biopsies, offering a window into the complex cellular interactions underlying gut function and dysfunction, to investigate their roles in health and disease [5, 7, 8]. This section serves as a foundational guide to the meticulous methodologies employed in this endeavor. This chapter presents a comprehensive protocol for EGC isolation, detailing the sequential procedures involved in obtaining a pure population of EGCs suitable for downstream applications. The outlined method encompasses a series of meticulously designed steps aimed at preserving the integrity and viability of EGCs extracted from gastrointestinal tissue. Beginning with tissue collection, close coordination with clinical endoscopy teams ensures the timely acquisition of mucosal biopsies while minimizing trauma to tissues. The subsequent enzymatic digestion process involves the use of specific enzymes, such as collagenase Type II and Dispase II, to break down the extracellular matrix and release cells for further analysis [9]. Mechanical dissection further refines the process, allowing for the uniform fragmentation of tissue while maintaining sterile conditions to prevent contamination [10]. Finally, cell purification steps ensure the removal of debris and aggregates, culminating in the isolation of viable EGCs ready for downstream applications.

The protocol outlined delineates four major procedures crucial to the successful isolation and dissection of EGCs:

1. Tissue Collection: Coordinated collection of human mucosal biopsies from targeted gastrointestinal regions, ensuring minimal trauma and optimal preservation during transportation.
2. Enzymatic Digestion: Utilization of specific enzyme solutions to break down the extracellular matrix and release cells from the tissue, with careful optimization of conditions for maximal yield and viability.
3. Mechanical Dissection: Precise mincing of tissue into uniform fragments using sterile surgical instruments, maintaining sterile conditions to preserve cell viability [11, 12].
4. Cell Purification: Filtration and centrifugation steps to remove debris and aggregates, resulting in the isolation of purified EGCs suitable for downstream applications [7]. By adhering to the outlined procedures, researchers and clinicians can isolate purified EGCs, shedding light on the intricate roles of EGCs in gastrointestinal health and disease.

2 Materials

2.1 Equipment

1. Antibiotic Supplement: Penicillin–streptomycin solution.
2. Tissue Culture Plates/Flasks: Use sterile tissue culture plates or flasks with appropriate surface coatings (e.g., tissue culture-treated, poly-D-lysine-coated) for cell culture experiments (*see* **Note 1**).
3. Centrifuge Tubes: Select sterile centrifuge tubes compatible with the centrifuge rotor and capable of withstanding high centrifugal forces (*see* **Note 2**).
4. Microscope Slides and Coverslips: Prepare clean microscope slides and coverslips for cell staining and microscopic analysis. Ensure that slides are free from dust, grease, or other contaminants that may interfere with imaging (*see* **Note 3**).

2.2 Buffers and Solutions

1. Sterile Surgical Instruments: Utilize autoclaved surgical instruments, including forceps, scissors, and scalpel blades (*see* Fig. 2) for tissue handling and dissection. Ensure that instruments are properly sterilized and stored in a clean, dry environment to prevent contamination (*see* **Note 4**).
2. Hanks' Balanced Salt Solution (HBSS): 8 g/L NaCl, 0.4 g/L KCl, 0.06 g/L Na_2HPO_4, 0.06 g/L KH_2PO_4, 0.1 g/L $MgCl_2$, 0.1 g/L $CaCl_2$, and 1 g/L glucose, in deionized or distilled water. Adjust the pH to 7.4 with HCl or NaOH as needed. Filter the solution using a 0.22 μm membrane filter and store at 4 °C.

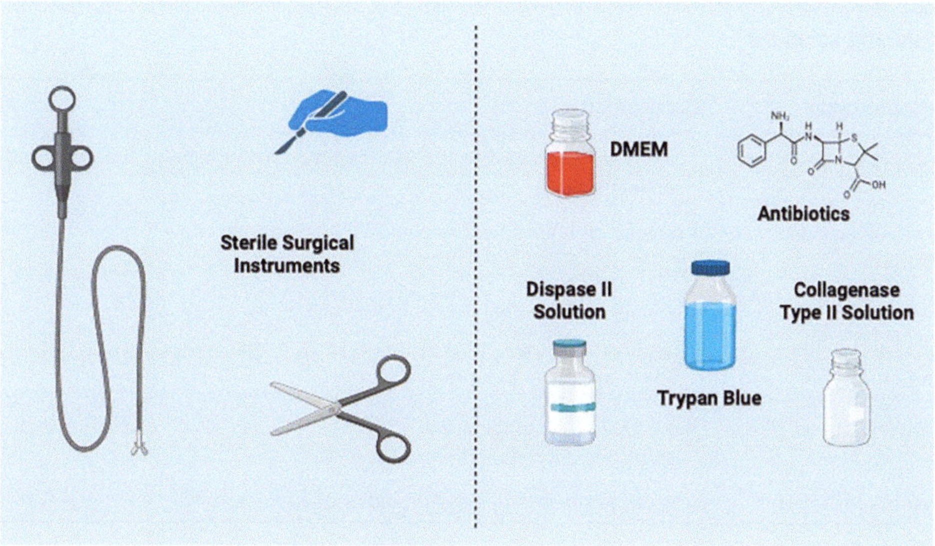

Fig. 2 Materials: instruments, medium, and solutions

3. Collagenase Type II Solution: 10 mg/mL in sterile, ice-cold HBSS. Aliquot the stock solution into smaller volumes and store at −20 °C for long-term use. Thaw aliquots immediately before use and discard any unused portion to prevent enzyme degradation.
4. Dispase II Solution: 5 mg/mL in sterile, ice-cold HBSS. Store aliquots at −20 °C and thaw as needed. Avoid repeated freeze–thaw cycles to maintain enzyme activity.
5. Trypan Blue Solution: Prepare a 0.4% (w/v) trypan blue in phosphate-buffered saline (PBS), using a 0.22 μm membrane filter and store at 4 °C protected from light; discard any unused solution after 1 week.

   See **Notes 5** and **6** for more.

2.3 Cell Culture Reagents

Fetal Bovine Serum (FBS): Source high-quality FBS from reputable suppliers to ensure consistent cell culture performance [13]. Heat-inactivate FBS by incubating at 56 °C for 30 min before use to inactivate complement proteins and reduce the risk of immune reactions.

Dulbecco's Modified Eagle Medium (DMEM): Basal DMEM with 10% (v/v) FBS and 1% (v/v) 100 U/mL penicillin and 100 μg/mL streptomycin (penicillin–streptomycin solution) [14]. Filter the medium using a 0.22 μm membrane filter and store at 4 °C until use.

See **Notes 7** and **8** and Table 2 for more.

Table 2
Materials and solutions

Material/Solution	Description
Sterile surgical instruments	Scissors, forceps, and scalpel blades autoclaved for tissue handling and dissection
Type II collagenase solution	Prepared at 10 mg/mL in cold HBSS. Stored at −20 °C
Dispase II solution	Prepared at 5 mg/mL in cold HBSS. Stored at −20 °C
Hanks' balanced salt solution (HBSS)	Composition: NaCl, KCl, Na2HPO4, KH2PO4, MgCl2, $CaCl_2$, glucose. pH 7.4. Filtered and stored at 4 °C
Culture medium (DMEM)	Basal DMEM supplemented with 10% FBS and 1% penicillin–streptomycin solution. Filtered and stored at 4 °C
Trypan blue solution	Prepared at 0.4% in PBS. Filtered and stored at 4 °C

3 Methods

3.1 Tissue Collection

1. Coordinate closely with clinical endoscopy teams (*see* Fig. 3) to ensure timely collection of human mucosal biopsies from targeted gastrointestinal regions [15]. Carefully collect human mucosal biopsies from the desired gastrointestinal region, ensuring minimal trauma to the tissue.
2. Utilize flexible endoscopes equipped with biopsy forceps or endoscopic biopsy tools to obtain tissue samples under direct visualization [16].
3. Minimize tissue handling time and temperature fluctuations by promptly transferring the biopsies to a sterile container filled with ice-cold Hanks' Balanced Salt Solution (HBSS) to maintain tissue integrity and viability during transportation to the laboratory.
4. Label the container with appropriate patient identifiers and tissue collection details for traceability (*see* **Notes 9** and **10**).

3.2 Enzymatic Digestion

1. Upon arrival in the laboratory (*see* **Note 11**), transfer the mucosal biopsies to a sterile Petri dish containing pre-warmed HBSS supplemented with collagenase Type II (1–2 mg/mL) and Dispase II (0.5–1 mg/mL).
2. Ensure a sufficient volume of enzyme solution to completely submerge the tissue. Optimize enzyme concentrations and digestion times (*see* **Note 12**), based on tissue type, age, and disease state, to achieve maximal cell yield and viability. Perform preliminary trials with varying enzyme concentrations and incubation times using surplus tissue samples to establish optimal digestion conditions [10].

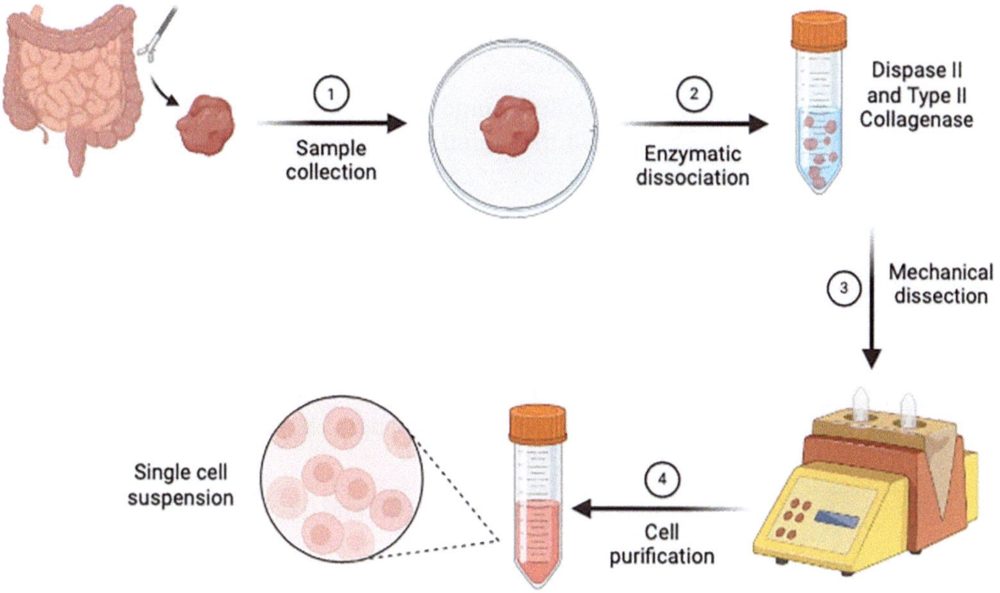

Fig. 3 Isolation and dissection procedure, in brief

3. Incubate the tissue in the enzyme solution at 37 °C for 20–30 min with gentle agitation using a magnetic stirrer or orbital shaker. Utilize mild agitation methods such as gentle rocking or orbital shaking to facilitate enzyme penetration and tissue dispersion without inducing mechanical stress on cells [9].
4. Monitor the digestion progress by visual inspection under a dissecting microscope, observing tissue fragmentation and detachment of cells from the extracellular matrix.

3.3 Mechanical Dissection

1. Following enzymatic digestion, transfer (*see* **Note 13**) the tissue to a new Petri dish containing fresh HBSS to halt the enzymatic activity.
2. Use sharp, sterile surgical instruments (e.g., forceps, fine scissors, razor blades) to finely mince tissue into uniform fragments (~1 mm), avoiding excessive mechanical disruption [12] (*see* **Note 14**).
3. Standardize tissue mincing techniques to ensure consistent cell dissociation and minimize sample heterogeneity (*see* **Note 15**).
4. Triturate the minced tissue gently with a Pasteur pipette to further dissociate cells into a homogeneous suspension (*see* **Note 16**).

3.4 Cell Purification

1. After mechanical dissociation (*see* **Note 17**), pass the cell suspension through a sterile 70 μm cell strainer placed over a clean collection tube.
2. Use a sterile plunger or pipette to gently press the tissue suspension through the mesh (*see* **Note 18**) of the cell strainer, facilitating the removal of large debris and aggregates.
3. Centrifuge (*see* **Note 19**) the filtered cell suspension at 300–400× *g* for 5 min at room temperature to pellet the cells. Use low-speed centrifugation (e.g., 100–300× *g*) to minimize mechanical stress on delicate cells and maximize recovery rates.
4. Aspirate supernatants to avoid disturbing cell pellets and resuspend cells gently in ice-cold HBSS or appropriate cell culture medium for downstream applications to maintain viability and prevent aggregation.

For more, *see* **Notes 20–23** and Tables 3 and 4.

Table 3
EGC isolation and dissection procedure

Procedure	Description
Tissue collection	Coordination with clinical teams to obtain mucosal biopsies with minimal trauma. Rapid transfer to cold HBSS solution
Enzymatic digestion	Use of enzymatic solutions (Type II collagenase and Dispase II) to break down the extracellular matrix and release cells
Mechanical dissection	Mince tissues into uniform fragments using sterile surgical instruments. Gentle trituration to disperse cells without introducing air bubbles
Cell purification	Filtration to remove debris, followed by centrifugation to pellet the cells. Resuspension in an appropriate medium to ensure cell viability

Table 4
Detailed methodology

Step	Description
1. Tissue collection	Collect biopsies using flexible endoscopes, rapidly transfer to containers with cold HBSS
2. Enzymatic digestion	Immerse tissues in pre-warmed HBSS with type II collagenase (1–2 mg/mL) and Dispase II (0.5–1 mg/mL) at 37 °C for 20–30 min with gentle agitation
3. Mechanical dissection	Transfer digested tissue to a new petri dish with fresh HBSS. Mince into uniform fragments (~1 mm). Gentle trituration with a Pasteur pipette
4. Cell purification	Filter cell suspension through a 70 μm strainer. Centrifuge at 300–400× *g* for 5 min. Resuspend cells for further applications

4 Notes

1. Tissue Culture Plates/Flasks: Choose vessel sizes based on the scale of cell culture experiments and seeding densities.
2. Centrifuge Tubes: Use tubes with conical bottoms for efficient pellet formation during cell centrifugation steps.
3. Microscope Slides and Coverslips: Ensure that slides are free from dust, grease, or other contaminants that may interfere with imaging.
4. Sterility: Maintain strict aseptic techniques throughout the experiment to prevent microbial contamination of cell cultures. Use sterile disposable gloves, work in a laminar flow hood, and disinfect work surfaces with appropriate disinfectants.
5. Quality Control: Regularly perform quality control checks on reagents, buffers, and equipment to ensure consistent experimental results. Test new lots of enzymes for activity and optimize concentrations accordingly.
6. Temperature Control: Use pre-warmed buffers and maintain consistent incubation temperatures as specified in the protocol.
7. Temperature Control: Maintain appropriate temperature conditions during tissue collection, enzymatic digestion, and cell culture to preserve cell viability and functionality.
8. Waste Disposal: Dispose of biological waste, including used culture medium, cell culture plates, and contaminated materials, according to institutional guidelines and regulations for biohazardous waste disposal.
9. Macroscopic Evaluation: Before enzymatic digestion, visually inspect tissue specimens under a dissecting microscope to assess gross morphology and identify any abnormalities (e.g., ulcerations, fibrotic areas). Exclude visibly damaged or necrotic tissue from further processing to prevent contamination and obtain reliable results.
10. Histological Analysis: Perform histological examination of tissue sections using hematoxylin and eosin (H&E) staining to evaluate tissue architecture, cellularity, and inflammation. Assess the presence of mucosal layers, glandular structures, and inflammatory infiltrates to ensure sample integrity and suitability for cell isolation.
11. Concentration Titration: Conduct preliminary experiments to determine the optimal concentrations of collagenase Type II and Dispase II for tissue digestion. Evaluate a range of enzyme concentrations (e.g., 0.5–2 mg/mL) to achieve efficient tissue dissociation while minimizing cell damage and preserving viability.

12. Incubation Time Optimization: Explore various digestion durations (e.g., 15–60 min) to identify the optimal incubation period for tissue digestion. Monitor cell dissociation kinetics over time and balance enzymatic activity with the need to minimize tissue degradation and maintain cell viability.
13. Tissue Handling Guidelines: Handle tissue samples gently to avoid mechanical stress and maintain cell viability. Use sterile instruments with fine tips to minimize tissue trauma during mincing and triturating steps. Minimize tissue manipulation time to reduce the risk of cell death or activation.
14. Optimal Fragment Size: Aim to achieve uniform tissue fragmentation into small, homogenous fragments (~1 mm) to facilitate enzyme penetration and maximize cell yield. Avoid over-mincing tissue, as excessive disruption may release intracellular contents and compromise cell viability.
15. Maintain a sterile work environment and aseptic conditions throughout the dissection process. Handle tissue samples with care to prevent contamination and maintain cell viability. Avoid excessive manipulation of the tissue to minimize cellular damage and preserve cell viability.
16. Perform this step with caution to avoid introducing air bubbles into the cell suspension, which may affect subsequent analyses.
17. Aseptic Technique: Adhere strictly to sterile techniques to prevent contamination and ensure the purity of isolated cell populations.
18. Filtration Efficiency: Select appropriate filtration methods (e.g., cell strainers, mesh sieves) based on cell size and sample volume to remove debris and aggregates while retaining viable cells. Validate the efficiency of cell strainers or sieves in removing debris and large aggregates from the cell suspension. Verify that selected pore sizes adequately retain intact cells while allowing passage of dissociated cells and smaller fragments.
19. Centrifugation Parameters: Optimize centrifugation conditions (e.g., speed, duration) to achieve efficient cell pelleting without excessive mechanical force. Avoid high-speed centrifugation, which may lead to cell damage or loss, particularly for delicate cell types such as enteric glial cells.
20. Sample Characterization: Validate isolated cell populations using immunocytochemistry, flow cytometry, or molecular assays to confirm cell identity and purity.
21. Trypan Blue Exclusion Assay: Perform trypan blue staining to assess cell viability before and after isolation procedures. Count viable and non-viable cells using a hemocytometer or automated cell counter to calculate cell viability percentages and ensure the quality of isolated cell populations.

22. Live/Dead Staining: Consider complementary viability assays such as calcein-AM/ethidium homodimer-1 staining for flow cytometric analysis or fluorescent microscopy. Evaluate cell membrane integrity and intracellular esterase activity to discriminate live cells from dead or apoptotic cells accurately.
23. Data Standardization: Document detailed protocol parameters and experimental conditions to facilitate reproducibility and comparison across studies.

References

1. Hoff S, Zeller F, von Weyhern CW et al (2008) Quantitative assessment of glial cells in the human and Guinea pig enteric nervous system with an anti-Sox8/9/10 antibody. J Comp Neurol 509(4):356–371. https://doi.org/10.1002/cne.21769
2. Cirillo C, Sarnelli G, Esposito G et al (2011) S100B protein in the gut: the evidence for enteroglial-sustained intestinal inflammation. World J Gastroenterol 17(10):1261–1266. https://doi.org/10.3748/wjg.v17.i10.1261
3. Gulbransen BD, Sharkey KA (2019) Novel functional roles for enteric glia in the gastrointestinal tract. Nat Rev Gastroenterol Hepatol 9(11):625–632. https://doi.org/10.1038/nrgastro.2012.138
4. Neunlist M, Van Landeghem L, Mahé MM et al (2013) The digestive neuronal-glial-epithelial unit: a new actor in gut health and disease. Nat Rev Gastroenterol Hepatol 10(2):90–100. https://doi.org/10.1038/nrgastro.2012.221
5. Boesmans W, Lasrado R, Vanden Berghe P et al (2015) (2015) heterogeneity and phenotypic plasticity of glial cells in the mammalian enteric nervous system. Glia 63(2):229–241. https://doi.org/10.1002/glia.22746
6. Kabouridis PS, Lasrado R, McCallum S et al (2015) The gut microbiota keeps enteric glial cells on the move; prospective roles of the gut epithelium and immune system. Gut Microbes 6:398–403. https://doi.org/10.1080/19490976.2015.1109767
7. Cirillo C, Sarnelli G, Turco F et al (2011) Proinflammatory stimuli activates human-derived enteroglial cells and induces autocrine nitric oxide production. Neurogastroenterol Motil 23(9):e372–e382. https://doi.org/10.1111/j.1365-2982.2011.01748.x
8. Seguella L, Gulbransen BD (2021) Enteric glial biology, intercellular signalling and roles in gastrointestinal disease. Nat Rev Gastroenterol Hepatol 18:571–587. https://doi.org/10.1038/s41575-021-00423-7
9. Reichard A, Asosingh K (2019) Best practices for preparing a single cell suspension from solid tissues for flow cytometry. Cytometry A 95(2):219–226. https://doi.org/10.1002/cyto.a.23690
10. Turco F, Sarnelli G, Cirillo C et al (2014) Enteroglial-derived S100B protein integrates bacteria-induced Toll-like receptor signalling in human enteric glial cells. Gut 63(1):105–115. https://doi.org/10.1136/gutjnl-2012-302090
11. Morral C, Ghinnagow R, Karakasheva T et al (2023) Isolation of epithelial and stromal cells from colon tissues in homeostasis and under inflammatory conditions. Bio Protoc 13(18):e4825. https://doi.org/10.21769/BioProtoc.4825
12. Caldwell ML, Cook CA, Mariant CL, Touvron M, Odle J, Blikslager AT, Ziegler AL, Van Landeghem L (2024) Protocol to enteric glial cells from the submucosal and myenteric plexi of neonatal and juvenile pig colons. STAR Protoc 5(2):103057. https://doi.org/10.1016/j.xpro.2024.103057
13. Freshney RI (2010) Culture of animal cells: a manual of basic technique and specialized applications, 6th edn. Wiley
14. Sambrook J, Fritsch EF, Maniatis T (1989) Molecular cloning: a laboratory manual, 2nd edn. Cold Spring Harbor Laboratory Press
15. ASGE Technology Committee, Thosani N (2017) Endoscopic mucosal tissue sampling. Gastrointest Endosc 85(1):25–33
16. ASGE Standards of Practice Committee, Fisher DA, Maple JT, Ben-Menachem T et al (2016) Complications of colonoscopy. Gastrointest Endosc 74(4):745–752

Chapter 7

Differentiating Enteric Glial Cells Within Intestinal Organoids Derived from Human Pluripotent Stem Cells

Simon Vales, Lola Bonneau, and Maxime M. Mahe

Abstract

Human intestinal organoids (HIOs) derived from pluripotent stem cells are valuable tools for studying gastrointestinal development and function. Herein, we present a method for generating and co-culturing human intestinal organoids and enteric glial cells derived from human pluripotent stem cells. In addition, we describe how to collect and stain enteric glial cells within organoids embedded in Matrigel. The method outlines the precise steps required to successfully generate both human intestinal organoids and vagal neural crest cells, offering valuable insight into the study of human enteric glial cells.

Key words Human intestinal organoids, Pluripotent stem cells, Enteric glial cells, Vagal neural crest cells, Stepwise differentiation

1 Introduction

The digestive tract is a complex organ with conflicting roles of absorbing fluids and nutrients while preventing the passage of pathogens. To accomplish these tasks, the gut possesses its own autonomous nervous system, known as the enteric nervous system (ENS). The ENS controls various functions such as peristaltic movements, nutrient assimilation, secretion of electrolytes and mucus, and regulation of blood flow [1]. Composed of hundreds of millions of neurons, the ENS has earned the moniker of "the second brain" [2]. Alongside enteric neurons, the ENS comprises glial cells, outnumbering neurons by 4–6 times depending on the species [3]. Enteric glial cells (EGCs) are non-excitable cells characterized by their irregular shape, small cell bodies, variable-sized processes, and expression of GFAP, S-100β, Sox10, and PLP1. Beyond providing support to neurons by acting on neuronal survival or neurotransmission [4–6], EGCs interact with multiple cell

Giuseppe Esposito (ed.), *Enteric Glial Cells: Methods and Protocols*, Methods in Molecular Biology, vol. 2971,
https://doi.org/10.1007/978-1-0716-4795-0_7,

types of the gut and regulate gastrointestinal functions, including motility and host defense [7, 8]. Moreover, EGCs have been described as an important player of the intestinal niche [9], and several studies have highlighted EGCs‘pivotal role in regulating epithelial cell homeostasis and regeneration. For instance, CGEs have been shown to inhibit epithelial cell proliferation [10], reinforce the intestinal barrier [11], and enhance epithelial healing [12, 13]. Besides being an important player in intestinal functions, EGCs are also involved in a number of digestive pathologies such as infection [14], inflammatory bowel disease [15], neuroinflammation [16] obesity [17, 18], colorectal cancer [19], idiopathic slow transit constipation [20], or in neurodegenerative diseases such as Parkinson’s disease [21].

Until now, EGCs have been studied using in vitro culture and co-culture models or animal models. Today, innovative cellular models can offer a new approach to the study of CGEs. These include intestinal organoids derived from pluripotent stem cells. Using a stepwise process that mimics embryonic intestinal development, human pluripotent stem cells (hPSC) can be differentiated into hindgut structures that develop into self-assembled multilayered organoids containing an epithelium, neurons, and glial cells embedded in a supporting mesenchyme [22, 23]. These human intestinal organoids (HIOs) resemble the original human tissue, contain the correct cell types, and recapitulate at least some functions of the organ [24]. Therefore, HIOs can provide an interesting alternative to in vivo models, enabling the in vitro study of glomerular epithelial cells (GECs) in a system integrating a wide range of cells, including neurons, mesenchyme, and intestinal epithelium, with a tissue organization similar to that of the human intestine.

Herein, we present a method for generating and co-culturing human intestinal organoids and enteric glial cells derived from human pluripotent stem cells. Three main steps include vagal neural crest cells (vNCC) differentiation, human intestinal organoid (HIO) differentiation, followed by HIO-vNCCs co-culture in order to generate an HIO+ENS. The overall process takes 41 days, including 13 days to differentiate vNCCs, 9 days to differentiate HIOs, and 28 days of HIO-vNCCs co-culture. Some parts of this manuscript are adapted from a previous protocol [25]. In addition, we describe how to collect and stain enteric glial cells within organoids embedded in Matrigel. The method highlights the specific steps to the successful generation of both human intestinal organoids and vagal neural crest cells and provides insight into the study of human enteric glial cells.

2 Materials

2.1 Vagal Neural Crest Cells (vNCC) Differentiation

1. AggreWell 800 plates.
2. Anti-adherence rinsing solution.
3. DMEM/F12, HEPES.
4. Accutase.
5. vNCC medium: 50% DMEM/F-12 HEPES, 50% Neurobasal Medium supplemented with 2 mM GlutaMAX, 1× B-27 and N2 supplements, Insulin (5 μg/mL), FG2 (20 ng/mL), EGF (50 ng/mL), and 100 U/mL penicillin–streptomycin.
6. ROCK specific inhibitor Y27632, working concentration 10 μM.
7. Retinoic acid, working concentration 2 μM.
8. Fibronectin, working concentration 1.5 μg/mL in sterile Dulbecco's phosphate-buffered saline (DPBS).
9. DPBS 1X
10. Nunclon Delta Surface 6-well plates.

2.2 Human Intestinal Organoid (HIO) Generation

1. mTESR-1.
2. DPBS 1X.
3. Matrigel hESC-Qualified Matrix.
4. ROCK inhibitor Y27632, working concentration 10 μM.
5. DMEM/F-12, HEPES.
6. Accutase.
7. Hindgut differentiation medium: Roswell park memorial institute (RPMI) Medium 1640 supplemented with 1× MEM non-essential amino acids solution, 0–2% heat-inactivated (HI) fetal bovine serum.
8. Activin A, working concentration 100 ng/mL.
9. Human recombinant fibroblast growth factor 4, working concentration 500 ng/mL.
10. Chiron, working concentration 3 μM.
11. Nunclon Delta 24-well plates.

2.3 Human Intestinal Organoid (HIO) Maintenance

1. Matrigel growth factor reduced.
2. Intestinal growth medium: Advanced DMEM/F12 medium supplemented with 2 mM glutamine, 10 mM HEPES, 100 U/mL penicillin, 100 μg/mL streptomycin, 1× N2 supplement, 1× B27 supplement, and recombinant epidermal growth factor (100 ng/mL).

2.4 Whole Mount Staining

1. Phosphate buffered saline (PBS) 1×.
2. Paraformaldehyde, 2% diluted in DPBS.
3. Permeabilization solution: PBS supplemented with 0.5% Triton X-100.
4. Blocking solution: PBS supplemented with 5% bovine serum albumin (BSA).
5. Antibody diluent solution: PBS supplemented with 0.1% Triton X-100 and 1% BSA.
6. 4′,6-Diamidino-2-phenylindole (DAPI).
7. Rapiclear 1.52.

3 Methods

hPSCs maintenance and passaging have been described in the following protocols [25, 26].

3.1 Vagal Neural Crest Cells (vNCC) Differentiation (Fig. 1)

The protocol outlined here will lead to the generation of vagal neural crest cells (vNCC) in adhesion cultures in a 6-well plate format.

Day 0 (Neurosphere Generation)

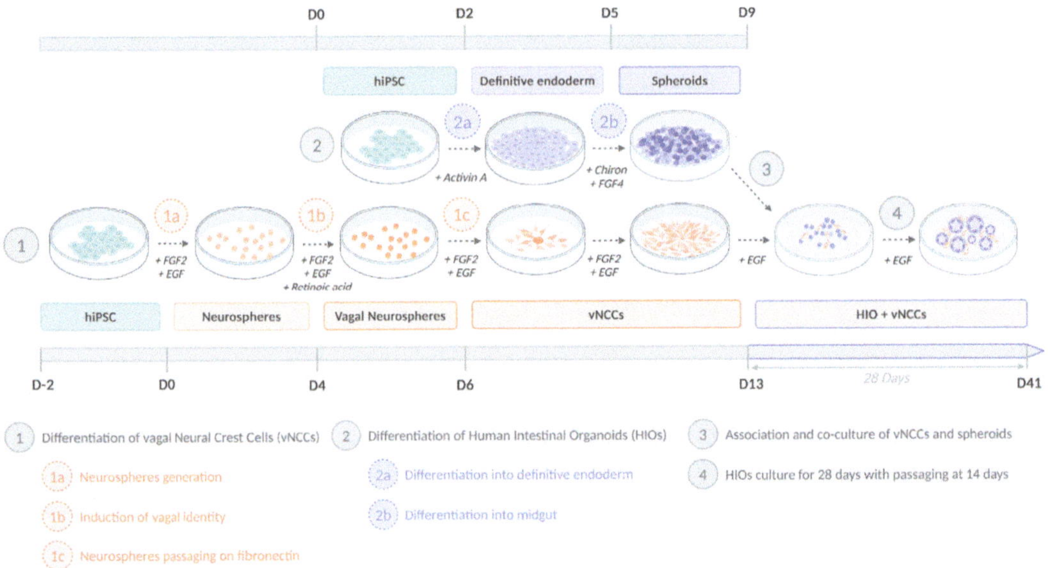

Fig. 1 Innervated human intestinal organoids differentiation protocol overview. The experiment starts with the vagal neural crest cells (vNCC) differentiation, followed by the human intestinal organoid generation (HIO). Association of both vNCCs and HIOs is performed at day 13. The combined cells result in the self-assembled intestinal organoid containing enteric glia and neurons (HIO+ENS)

3.1.1 Generation of PSC-Derived Neurospheres (Days 0–6)

1. Add 500 μL of anti-adherence rinsing solution to the AggreWell wells that will be used to generate the vagal neural crest cells (VNCCs) (*see* **Note 1**).
2. Centrifuge 5 min at 1300 g.
3. Rinse wells with 2 mL of DMEM/F12, HEPES.
4. Aspirate, add 1 mL of vNCC medium with ROCKi, and keep the plate aside.
5. Wash hPSCs well with 2 mL of DMEM/F12, HEPES at 37 °C.
6. Aspirate DMEM/F12.
7. Add 1 mL of Accutase prewarmed at 37 °C.
8. Incubate the plate for 5 min at 37 °C, 5% CO_2 (*see* **Note 2**).
9. Confirm detachment of the colonies' border under a microscope.
10. Add 2 mL of DMEM/F12, HEPES.
11. Pipette back and forth to dissociate cell clumps into single cells.
12. Collect the cells into a 15 mL conical tube.
13. Centrifuge at 300 g for 1 min.
14. Aspirate the supernatant without disturbing the pellet.
15. Resuspend the pellet with 1 mL of vNCC medium with ROCKi.
16. Count cells using a Malassez counting chamber or other cell counting methods.
17. Calculate the appropriate volume of cell solution to add 1,200,000 cells to each AggreWell well prepared in **steps 1–4**.
18. Complete volume up to 2 mL in each AggreWell well with vNCC medium with ROCKi.
19. Centrifuge AggreWell plate at 100 g for 3 min.
20. Incubate AggreWell plate in the incubator at 37 °C, 5% CO_2.

Days 2–4 (Daily Medium Change)

1. Gently remove vNCC medium from the wells without aspirating the spheres (*see* **Note 3**).
2. Gently add 2 mL of vNCC medium.
3. Return the plate to the incubator at 37 °C, 5% CO_2.
4. Repeat **steps 1–3** every day until day 4.

Days 4–6 (Induction of Vagal Identity)

1. Gently remove vNCC medium from the wells without aspirating the spheres.
2. Gently add 2 mL of vNCC medium supplemented with 2 μM of retinoic acid.

3. Return the plate to the incubator at 37 °C, 5% CO_2.
4. Repeat **steps 1–3** the next day.

3.1.2 Passaging of Neurospheres on Fibronectin-Coated Plates (Days 6–13)

vNCC culture requires the pre-coating of tissue culture plates with fibronectin. Thus, fibronectin-coated 6-well plates must be prepared at least an hour prior to passaging using 1.5 μg/mL fibronectin in sterile DPBS. 1 AggreWell well is sufficient for an entire 6-well plate. The pre-coated plates can be stored at 4 °C for a maximum of 3 days.

Days 6–10

1. Flush the medium from the AggreWell well to suspend the spheres. Then, aspirate the medium and transfer it into a 15 mL conical tube.
2. Wait for the spheres to settle in the tube.
3. Aspirate the supernatant without disturbing the pellet.
4. Gently resuspend 25–50 neurospheres in 1.5 mL of vNCC medium (*see* **Note 1**).
5. Aspirate fibronectin solution from the recipient 6-well plate and seed the neurospheres using a final volume of 1.5 mL of vNCC medium per well.
6. Shake the plate back and forth to evenly distribute the neurospheres in the well.
7. Return the plate to the incubator at 37 °C, 5% CO_2 (*see* **Note 4**).
8. Change the medium every day using 1.5 mL of vNCC medium per well until day 9.

Days 10–13 (Passaging of Neural Crest Cells)

Fibronectin-coated 6-well plates must be prepared at least an hour prior to passaging using 1.5 μg/mL fibronectin in sterile DPBS.

1. Aspirate vNCC medium.
2. Wash with 1 mL of DMEM/F12, HEPES.
3. Add 1 mL of Accutase.
4. Incubate 2 min at 37 °C (*see* **Note 5**).
5. Add 2 mL of DMEM/F12.
6. Transfer the cell suspension into a 15 mL conical tube.
7. Centrifuge at 300 g for 5 min.
8. Aspirate the supernatant without disturbing the pellet.
9. Resuspend the cell pellet in 1 mL of vNCC medium.

10. Count cells using a Malassez counting chamber or other cell counting methods.
11. Aspirate fibronectin solution from the recipient 6-well plate and seed 180,000 single cells per well in 1.5 mL of vNCC medium.
12. Shake the plate back and forth evenly to distribute the cell suspension in the well.
13. Return the plate to the incubator at 37 °C, 5% CO_2.
14. Change the vNCC medium every day with 1.5 mL per well until day 13.

3.1.3 Beyond Day 13 (Passaging of Neural Crest Cells)

vNCCs can be passaged every 3–4 days. Thus, we repeat the passaging of the neural crest cells from **steps 1–13**. vNCCs maintain their phenotype up to three passages before combination with intestinal spheroids. Although we do not typically use it, the vNCCs from the first or second passages can be frozen and thawed out using standard techniques.

3.2 HIO Generation (Fig. 1)

The protocol outlined here describes the generation of human hindgut spheroids emerging from adhesion cultures in a 24-well plate format. HIO differentiation requires the pre-coating of 24-well plates with hESC-qualified Matrigel diluted in DMEM/F12 medium. The dilution factor is provided on the Certificate of Analysis for each Matrigel lot. 300 μL of diluted hESC-qualified Matrigel must be added in each well of the 24-well plate and incubated at 37 °C minimum of 2 h and up to 24 h before use.

Days 0–2 (hPSC Passaging and Plating)

1. Aspirate mTeSR1 from hPSC wells.
2. Wash with 1 mL of DPBS per well.
3. Aspirate DPBS.
4. Add 1 mL of Accutase to each well.
5. Incubate 5 min at 37 °C (*see* **Note 2**).
6. Add 2 mL of DMEM/F12, HEPES.
7. Pipette back and forth to dissociate cell clumps into single cells.
8. Collect the cell suspension from the wells in a 15 mL conical tube (*see* **Note 6**).
9. Centrifuge for 1 min at 300 g.
10. Aspirate the supernatant.
11. Resuspend the cell pellet with 1 mL of mTESR-1 with ROCKi.
12. Count cells using a Mallasez chamber and calculate the volume needed of cell suspension to seed 24 wells at 100,000 cells per well (*see* **Note 7**).

13. Add the calculated volume of cell suspension in a 15 mL conical tube and complete volume up to 12 mL with mTESR-1 with ROCKi.
14. Distribute 500 μL of cell suspension per well of a 24-well plate.
15. Shake the plate back and forth to distribute the cells evenly in the wells.
16. Return the plate to the incubator at 37 °C, 5% CO_2 for 24 h.
17. Aspirate the medium.
18. Add 500 μL of mTSER-1 per well.
19. Return the plate to the incubator at 37 °C, 5% CO_2 for 24 h.

Days 2–5 (Definitive Endoderm Differentiation)

1. Aspirate medium on day 2.
2. Add 500 μL of Hindgut differentiation medium without HI FBS, supplemented with 100 ng/mL Activin A per well.
3. Return the plate to the incubator at 37 °C, 5% CO_2 for 24 h.
4. Aspirate medium on day 3.
5. Add 500 μL of Hindgut differentiation medium with 0.2% HI FBS supplemented with 100 ng/mL Activin A per well.
6. Return the plate to the incubator at 37 °C, 5% CO_2 for 24 h.
7. Aspirate the medium on day 4.
8. Add 500 μL of Hindgut differentiation medium with 2% HI FBS, supplemented with 100 ng/mL Activin A per well.
9. Return the plate to the incubator at 37 °C, 5% CO_2 for 24 h.

Days 5–9 (Intestinal Tissue Differentiation)

1. Aspirate medium.
2. Add 500 μL of Hindgut differentiation medium with 2% HI FBS, complemented with 500 ng/mL FGF4 and 3 μM Chiron.
3. Return the plate to the incubator at 37 °C, 5% CO_2 for 24 h.
4. Repeat **steps 1–3** for four consecutive days.

3.3 Generation of HIO + ENS

When spheroids emerge by day 9, vNCCs can be combined prior to their inclusion in an extracellular matrix gel (Matrigel).

3.3.1 Recovery of Intestinal Spheroids

1. Recover spheroids floating in the medium using a 1 mL micropipette.
2. Collect the spheroids in a 15 mL conical tube. Multiple wells from a 24-well plate can be combined into one tube.
3. Wait for the spheroids to settle in the tube at room temperature (*see* **Note 8**).
4. Aspirate out the medium.

5. Add 1 mL of intestinal growth medium.
6. Pipette 10 μL of solution and place it in a Petri dish to count the number of spheroids in order to estimate their total number in the tube.
7. Place the 15 mL conical tube in an incubator and proceed to the recovery of vNCCs.

3.3.2 Recovery of vNCCs

A well of a 6-well plate of vNCCs usually contains around $0.4–0.5 \times 10^6$ cells. We recommend using a ratio of 1000 vNCCs per spheroid.

1. Aspirate the medium.
2. Wash with 1 mL of DMEM/F12.
3. Add 1 mL of Accutase.
4. Incubate 2–3 min at 37 °C (*see* **Note 9**).
5. Add 5 mL of DMEM/F12.
6. Transfer the cell suspension to a 15 mL conical tube.
7. Centrifuge at 300 g for 5 min.
8. Aspirate supernatant without disturbing the pellet.
9. Resuspend the cell pellet in 1 mL of DMEM/F12.
10. Count cells using a Malassez counting chamber or other cell counting methods.
11. Proceed to the co-culture of intestinal spheroids and vNCCs.

3.3.3 Co-Culture of Intestinal Spheroids and vNCCs (Fig. 2)

1. Add the required volume of vNCC suspension to the 15 mL conical tube containing the spheroids (*see* **Note 10**).
2. Centrifuge 2 min at 300 g.
3. Aspirate supernatant.

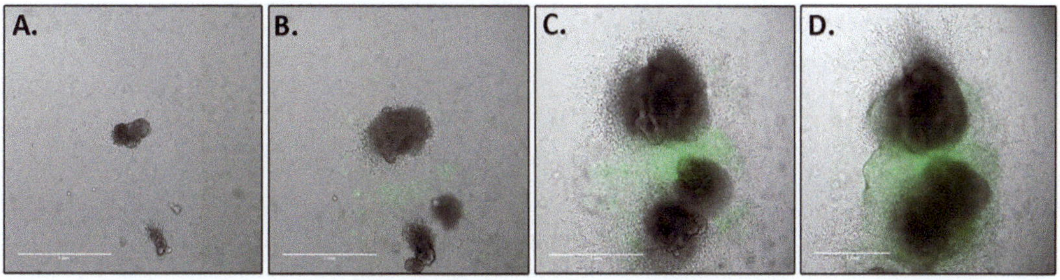

Fig. 2 Growth of innervated intestinal organoid in Matrigel over 13 days (from Loffet et al. [25]). GFP-expressing vagal neural crest cell differentiation (vNCC) is shown in green. (**a**) Innervated intestinal organoid (HIO+ENS) a day after the combination of hindgut spheroids and vNCCs. (**b**) HIO+ENS after 5 days in culture. (**c**) HIO+ENS after 9 days in culture. (**d**) HIO+ENS after 13 days in culture prior to splitting (scale bar: 1 mm)

4. Gently resuspend the spheroids and vNCCs in ice-cold Matrigel, pipetting up and down to evenly mix vNCCs with spheroids (*see* **Note 11**).
5. Distribute 20 μL droplets of ice-cold Matrigel (without vNCCs +spheroids) in the center of a 24-well plate (*see* **Note 12**).
6. Overlay the existing Matrigel drop with 20 μL of hindgut spheroids+vNCCs Matrigel suspension.
7. Allow Matrigel to start polymerization for 2–3 min at room temperature.
8. Turn the cell culture plate upside-down.
9. Return the plate to the incubator at 37 °C, 5% CO_2 for 20–30 min.
10. Overlay the Matrigel drop with 500 μL of intestinal growth medium.
11. Replace the medium twice a week for 14 days.

3.3.4 Passaging of Innervated Intestinal Organoids

Organoids should be passaged after 2 weeks of culture and placed in fresh Matrigel.

1. Collect the Matrigel drop containing HIO+ENS with a 1 mL micropipette under a stereomicroscope using sterile culture conditions (*see* **Note 13**). Transfer HIO+ENS in a sterile Petri dish filled with DMEM/F12, HEPES.
2. Under a stereomicroscope, separate HIO+ENS from remnant Matrigel using fine forceps.
3. If multiple HIO+ENS structures appear to be fused together, separate them into individual structures using fine forceps (*see* **Note 14**).
4. Collect HIO+ENS in a 15 mL conical tube.
5. Allow HIO+ENS to settle to the bottom of the tube at room temperature.
6. Aspirate supernatant.
7. Resuspend HIO+ENS in ice-cold Matrigel (*see* **Note 15**).
8. Distribute 35 μL of Matrigel containing HIO+ENS in the center of the well of a 24-well plate.
9. Allow Matrigel to start polymerization for 2–3 min at room temperature.
10. Turn the cell culture plate upside-down.
11. Return the plate to the incubator at 37 °C, 5% CO_2 for 20–30 min.
12. Overlay the Matrigel drop with 500 μL of intestinal growth medium.
13. Replace the medium twice a week for 14 days.

3.4 Differentiation Benchmarking

Each step in the differentiation protocol can be monitored using a combination of gene or/and protein expression. For example, definitive endoderm can be assessed by controlling the proportion of CXCR4-expressing cells by flow cytometry or FOXA2 and SOX17-positive cells using immunofluorescence. Similarly, the expression of CDX2 can be verified in hindgut spheroids. Finally, the expression of CD49d, HNK-1, and $P75^{NTR}$ can be tested in vNCCs.

3.5 Analysis—Fluorescent Immunostaining (Fig. 3)

The following protocol describes wholemount immunofluorescence staining on HIOs in Matrigel.

3.5.1 Organoid Fixation

1. Aspirate the medium and add 500 μL of a 2% paraformaldehyde (PFA) solution to each well.
2. Incubate for 1 h at room temperature.
3. Aspirate the PFA solution.
4. Wash with 500 μL of PBS for at least 10 min.
5. Repeat PBS wash 3 times (*see* **Note 16**).

3.5.2 Wholemount Organoid Staining

1. Permeabilize HIOs for 30 min with 500 μL of permeabilization solution on a rocking platform.
2. Remove permeabilization solution and add 500 μL of blocking solution for 1 h on a rocking platform.
3. Remove blocking solution and incubate primary antibodies diluted in antibody solution overnight at 4 °C on a rocking platform (Table 1).
4. Remove the antibody solution and wash with 500 μL of PBS on a rocking platform for at least 10 min. Repeat the procedure three times.

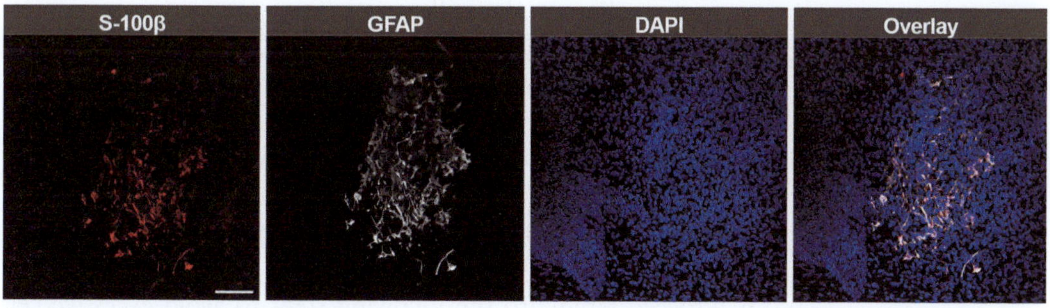

Fig. 3 Immunostaining of enteric glial cells in HIO + ENS. Immunofluorescence staining for the glial markers S-100β (red) and GFAP (white) showing their presence in HIOs with ENS in vitro (scale bar: 100 μm)

Table 1
Antibodies

	Antigen	Host	Dilution
Primary	S-100β	Guinea pig	1:400
	GFAP	Rabbit	1:400
Secondary	Guinea pig	Donkey	1:500
	Rabbit	Donkey	1:500

5. Remove PBS and incubate secondary antibodies diluted in antibody solution for 1 h at room temperature on a rocking platform (Table 1).
6. Remove the antibody solution and wash with 500 μL of PBS on a rocking platform for at least 10 min. Repeat the procedure three times.
7. Remove PBS and incubate with 1 μg/mL of DAPI in PBS for 10 min.
8. Remove DAPI and wash with PBS for 10 min.
9. Remove PBS and add 300 μL of Rapiclear 1.52. Incubate at 4 ° C overnight, and proceed to imaging using a confocal microscope (*see* **Note 17**).

4 Notes

1. Each AggreWell well should generate 300 spheres.
2. Incubation time can vary depending on the enzymatic activity of Accutase and the confluency of the well. Accusate must be diluted as soon as the cell clumps are detached from the bottom of the plate to prevent excessive cell death.
3. To limit the risk of spheres being sucked up, use a P1000 to aspirate the medium in two stages, placing the pipette tip against the wall of the well. Use the same strategy to add fresh medium to the well.
4. Neurospheres will adhere to the fibronectin-coated plate and form rosettes.
5. Confirm cell detachment under an inverted microscope. Only migrating vagal neural crest cells are picked up; the rosettes remain attached to the plate. We recommend a 2 min incubation with Accutase. Longer treatment will result in a detachment of the rosettes.
6. Proceed rapidly to the following steps to avoid excessive cell death, as the enzyme Accutase is just diluted and not inactivated.

7. Optimal conditions for HIO generation may vary depending on the pluripotent stem cell lines. The starting cell density is a critical aspect that needs to be tested for every hPSC line. A reasonable starting point is 100,000 cells per well of a 24-well plate. However, we recommend testing several densities ranging from 80,000 to 140,000 cells for each differentiation. The success of differentiation can be assessed by the hindgut spheroid yield.
8. Alternatively, you can centrifuge 1 min at 50 g to pellet spheroids.
9. Dilute with DMEM/F12 medium as soon as the cells are detached from the well to avoid excessive cell death.
10. We recommend using a ratio of 1000 vNCCs per spheroid.
11. We recommend embedding 20–40 spheroids per 20 μL of Matrigel. Matrigel should be kept on ice at all times to avoid polymerization.
12. We realize a Matrigel bed first to avoid vNCC accumulation at the bottom of the plate.
13. 1 mL pipette tips can be cut to facilitate HIO+ENS removal from the Matrigel
14. Under the stereomicroscope, single HIO+ENS can be selected and separated by identifying a central epithelial core.
15. We recommend embedding 5–10 HIO+ENS per 35 μL of Matrigel.
16. Fixed HIOs can be stored in PBS NaN3 at 4 °C for a few months.
17. HIOs can be imaged directly on the 24-well plate for low magnification. For high magnification, Matrigel drops can be carefully detached from the plate and mounted between a slide and a cover slip using a 0.25-mm iSpacer (SunJin lab).

References

1. Goyal RK, Hirano I (1996) The enteric nervous system. N Engl J Med 334(17): 1106–1115. https://doi.org/10.1056/NEJM199604253341707. Erratum in: N Engl J Med 1996;335(3):215
2. Gershon MD (1995) The enteric nervous system: a second brain. Hosp Pract 34(7):31–32, 35–38, 41–42 passim. https://doi.org/10.3810/hp.1999.07.153
3. Hoff S, Zeller F, von Weyhern CWH et al (2008) Quantitative assessment of glial cells in the human and guinea pig enteric nervous system with an anti-Sox8/9/10 antibody. J Comp Neurol. 509(4):356–371. https://doi.org/10.1002/cne.21769
4. Abdo H, Derkinderen P, Gomes P et al (2010) Enteric glial cells protect neurons from oxidative stress in part via reduced glutathione. FASEB J 24(4):1082–1094. https://doi.org/10.1096/fj.09-139519
5. Thomasi B, Gulbransen B (2023) Mini-review: intercellular communication between enteric glia and neurons. Neurosci Lett 806:137263. https://doi.org/10.1016/j.neulet.2023.137263

6. Brown IAM, Gulbransen BD (2018) The antioxidant glutathione protects against enteric neuron death in situ, but its depletion is protective during colitis. Am J Physiol Gastrointest Liver Physiol 314(1):G39–G52. https://doi.org/10.1152/ajpgi.00165.2017
7. Baghdadi MB, Kim TH (2023) The multiple roles of enteric glial cells in intestinal homeostasis and regeneration. Semin Cell Dev Biol 150–151:43–49. https://doi.org/10.1016/j.semcdb.2023.01.005
8. Liu C, Yang J (2022) Enteric glial cells in immunological disorders of the gut. Front Cell Neurosci 28(16):895871. https://doi.org/10.3389/fncel.2022.895871
9. Baghdadi MB, Ayyaz A, Coquenlorge S et al (2022) Enteric glial cell heterogeneity regulates intestinal stem cell niches. Cell Stem Cell 29(1):86–100.e6. https://doi.org/10.1016/j.stem.2021.10.004
10. Neunlist M, Aubert P, Bonnaud S et al (2006) Enteric glia inhibit intestinal epithelial cell proliferation partly through a TGF-beta1-dependent pathway. AJP: Gastrointest Liver Physiol 292(1):G231–G241. https://doi.org/10.1152/ajpgi.00276.2005
11. Savidge TC, Newman P, Pothoulakis C et al (2007) Enteric glia regulate intestinal barrier function and inflammation via release of S-Nitrosoglutathione. Gastroenterology 132(4):1344–1358. https://doi.org/10.1053/j.gastro.2007.01.051
12. Coquenlorge S, Van Landeghem L, Jaulin J et al (2016) The arachidonic acid metabolite 11β-ProstaglandinF2α controls intestinal epithelial healing: deficiency in patients with Crohn's disease. Sci Rep 6:25203. https://doi.org/10.1038/srep25203
13. Van Landeghem L, Chevalier J, Mahe MM et al (2011) Enteric glia promote intestinal mucosal healing via activation of focal adhesion kinase and release of proEGF. AJP: Gastrointest Liver Physiol 300(6):G976–G987. https://doi.org/10.1152/ajpgi.00427.2010
14. Selgrad M, De Giorgio R, Fini L et al (2009) JC virus infects the enteric glia of patients with chronic idiopathic intestinal pseudo-obstruction. Gut 58(1):25–32. https://doi.org/10.1136/gut.2008.152512
15. Le Berre C, Naveilhan P, Rolli-Derkinderen M (2023) Enteric glia at center stage of inflammatory bowel disease. Neurosci Lett 809:137315. https://doi.org/10.1016/j.neulet.2023.137315
16. Gaudette L, Gulbransen BD (2018) Communication between enteric neurons, glia, and nociceptors underlies the effects of tachykinins on neuroinflammation. Cell Mol Gastroenterol Hepatol 6(3):321–344. https://doi.org/10.1016/j.jcmgh.2018.05.009
17. Antonioli L, D'Antongiovanni V, Pellegrini C et al (2020) Colonic dysmotility associated with high-fat diet-induced obesity: role of enteric glia. FASEB J 34(4):5512–5524. https://doi.org/10.1096/fj.201901844R
18. Stenkamp-Strahm C, Patterson S, Boren J et al (2013) High-fat diet and age-dependent effects on enteric glial cell populations of mouse small intestine. Auton Neurosci 177(2):199–210. https://doi.org/10.1016/j.autneu.2013.04.014
19. Valès S, Bacola G, Biraud M et al (2019) Tumor cells hijack enteric glia to activate colon cancer stem cells and stimulate tumorigenesis. EBioMedicine 49:172–188. https://doi.org/10.1016/j.ebiom.2019.09.045
20. Bassotti G, Villanacci V, Maurer CA et al (2006) The role of glial cells and apoptosis of enteric neurones in the neuropathology of intractable slow transit constipation. Gut 55(1):41–46. https://doi.org/10.1136/gut.2005.073197
21. Clairembault T, Leclair-Visonneau L, Neunlist M et al (2015) Enteric glial cells: new players in Parkinson's disease? Mov Disord 30(4):494–498. https://doi.org/10.1002/mds.25979
22. Spence JR, Mayhew CN, Rankin SA et al (2011) Directed differentiation of human pluripotent stem cells into intestinal tissue in vitro. Nature 470(7332):105–109. https://doi.org/10.1038/nature09691
23. Workman MJ, Mahe MM, Trisno S et al (2017) Engineered human pluripotent-stem-cell-derived intestinal tissues with a functional enteric nervous system. Nat Med 23(1):49–59. https://doi.org/10.1038/nm.4233
24. Takebe T, Wells JM (2019) Organoids by design. Science 364(6444):956–959. https://doi.org/10.1126/science.aaw7567
25. Loffet E, Brossard L, Mahe MM (2020) Pluripotent stem cell derived intestinal organoids with an enteric nervous system. In: Methods in cell biology. Elsevier, pp 175–199. [cited 2020 Jul 1]. Available from: https://linkinghub.elsevier.com/retrieve/pii/S0091679X20300765
26. McCracken KW, Howell JC, Wells JM et al (2011) Generating human intestinal tissue from pluripotent stem cells in vitro. Nat Protoc 6(12):1920–1928. https://doi.org/10.1038/nprot.2011.410

Chapter 8

Transgenic Mouse Models to Study Enteric Glia

Wilmarie Morales-Soto

Abstract

In recent years, we have seen an explosion of research focused on understanding the intricate mechanisms revolving around enteric glia biology and their role in regulating gastrointestinal function and pathophysiology. This uptake in research has resulted in the development of novel technologies and approaches to study enteric glia. Numerous transgenic animal models have been developed that allow for the selective manipulation of enteric glia in vivo and ex vivo. In this chapter, we will review two transgenic animal models commonly used in glia research and provide detailed methods to implement them experimentally.

Key words Enteric glia, hM3Dq, DREADs, Inducible Cre recombinase, Tamoxifen

1 Introduction

Enteric glia are a large population of peripheral neuroglia that are associated with the cell bodies and processes of enteric neurons throughout the digestive tract [1]. Enteric glia play a critical role in intrinsic and extrinsic signaling in the gastrointestinal tract, but parsing out their specific contributions to these mechanisms has proven technically challenging. Transgenic mouse models have revolutionized biomedical research, offering powerful tools to investigate the intricate functions of various cell types, including enteric glia [2]. The evolution of these technologies within the last few decades has provided much insight into the specific contributions of enteric glia to homeostatic and pathophysiological processes.

Transgenic mouse models enable precise manipulation of enteric glia at the cellular level. For example, employing cell-specific promoters such as glial fibrillary acidic protein (GFAP), SRY-box transcription factor 10 (Sox10), proteolipid protein 1 (Plp1), or S100 calcium binding protein β (S100β) to drive the expression of Cre recombinase can be used to selectively targeted enteric glia for genetic modification [3–11]. The Cre-lox recombination system

Giuseppe Esposito (ed.), *Enteric Glial Cells: Methods and Protocols*, Methods in Molecular Biology, vol. 2971, https://doi.org/10.1007/978-1-0716-4795-0_8,

serves as a foundation in generating transgenic mouse models targeting enteric glia. Through this system, mice expressing Cre recombinase under glial-specific promoters can be crossed with mice harboring loxP-flanked genes of interest to induce gene deletion and overexpression or conditional activation. These techniques can be incorporated to elucidate the molecular mechanisms underlying enteric glial development, function, and response to pathological stimuli. Furthermore, Cre-lox technology can further our understanding of glia morphology, dynamics, and interactions with neighboring cell types [8, 9, 12–15]. Such approaches provide essential insights into the spatial organization and functional heterogeneity of enteric glia within the complex microenvironment of the gut.

This technology, however, is not without limitations. To date, there are no promoters that drive Cre expression exclusively in enteric glia. Cre expression under the control of the Sox10 and Plp1 promoters can also drive expression in other types of central and peripheral glia, like Schwann cells, satellite glia, and oligodendrocytes [16, 17]. Similarly, the GFAP promoter can drive Cre expression in astrocytes [18]. Many of these off-target effects are bypassed in ex vivo studies where the intestine is isolated. However, contributions from other glial cells should be taken into consideration when interpreting in vivo studies with these transgenic animal models. Nevertheless, transgenic mouse models have significantly advanced our understanding of enteric glial involvement in various gastrointestinal pathologies, including inflammatory bowel diseases, motility disorders, and visceral hypersensitivity [6, 8, 9, 12, 14]. Employing transgenic approaches can recapitulate disease-associated mutations or perturbations specifically within enteric glia, elucidating their contributions to disease pathogenesis and identifying potential therapeutic targets.

This chapter will discuss two transgenic mouse models used for the conditional activation of glia using chemogenetics (*GFAP: hM3Dq*) and the inducible deletion of glia-specific genes to disrupt enteric glia function (*Sox10-Cre*^{ERT2}). This chapter will also provide detailed methods to implement the use of these models.

1.1 Selective Activation of Glia Using Chemogenetics

Designer receptors exclusively activated by designer drugs (DREADDs) are powerful chemogenetic tools to manipulate cell-specific activity through the Gi-, Gq-, or Gs-protein-coupled signaling pathways [19]. DREADDs are used to elucidate glia functions, including those related to behavior and motility. The most widely applied DREADDs are based on human muscarinic receptors engineered to lose their affinity for acetylcholine and gain responsiveness to the artificial ligand clozapine-N-oxide (CNO), a metabolite of the atypical antipsychotic clozapine. The *GFAP: hM3Dq* [6, 20] is a constitutively derived transgenic mouse model containing the hemagglutinin (HA)-tagged hM3Dq driven by the

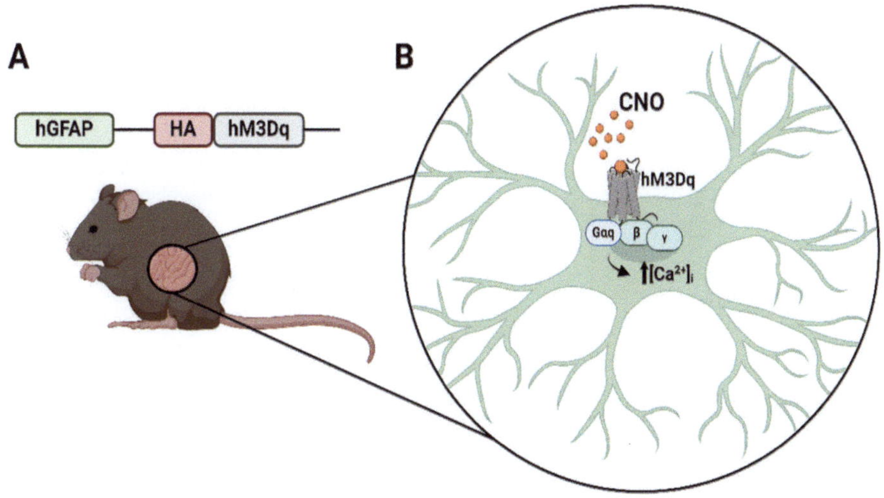

Fig. 1 Expression of hM3Dq receptors by enteric glia in the intestine of *GFAP::hM3Dq* mice. The schematic in Panel A depicts the transgene containing HA-tagged hM3Dq driven by the GFAP promoter. Panel B shows an example of hM3Dq signaling after CNO activation of enteric glia

GFAP promoter (Fig. 1). The selective manipulation of enteric glia through hM3Dq is particularly advantageous, as these cells naturally signal through Gq muscarinic receptors, replicating native signaling in enteric glia [7]. Furthermore, because this is a constitutive model, it can be bred with other inducible models to study neuro-glial interactions [8, 13, 21]. This mouse model has been readily used to study mechanisms of enteric glia in enteric nervous system signaling and gastrointestinal (GI) function in health and disease.

1.2 Selective Ablation of Glial Function Using Inducible Cre Transgenic Animal Models

Tamoxifen-inducible Cre-loxP is one of the most widely used inducible models for gene regulation in mice, because it allows spatial–temporal gene control. The inducible Cre-loxP estrogen receptor transgenic system is used to study cell-specific gene function through expression of Cre recombinase fused to a mutant form of the estrogen receptor (ER^{T2}) to create a $CreER^{T2}$ fusion protein. The ER^{T2} receptor does not bind its natural ligand (17β-estradiol), but it is activated when bound to the functional tamoxifen metabolite 4-hydroxytamoxifen (4-OHT). This approach circumvents the difficulties encountered when a Cre transgene is transiently expressed early in development.

Specifically, for enteric glia, the $Sox10^{CreERT2}$ [11] transgenic model serves as a powerful tool for selective genetic manipulation (Fig. 2). These mice can be crossed with mice containing a loxP-flanked sequence of interest to drive inducible Cre-mediated recombination. In the absence of tamoxifen, $CreER^{T2}$ is restricted to the cytoplasm. When tamoxifen or its metabolites are present, $CreER^{T2}$ is translocated into the nucleus, where it excises target

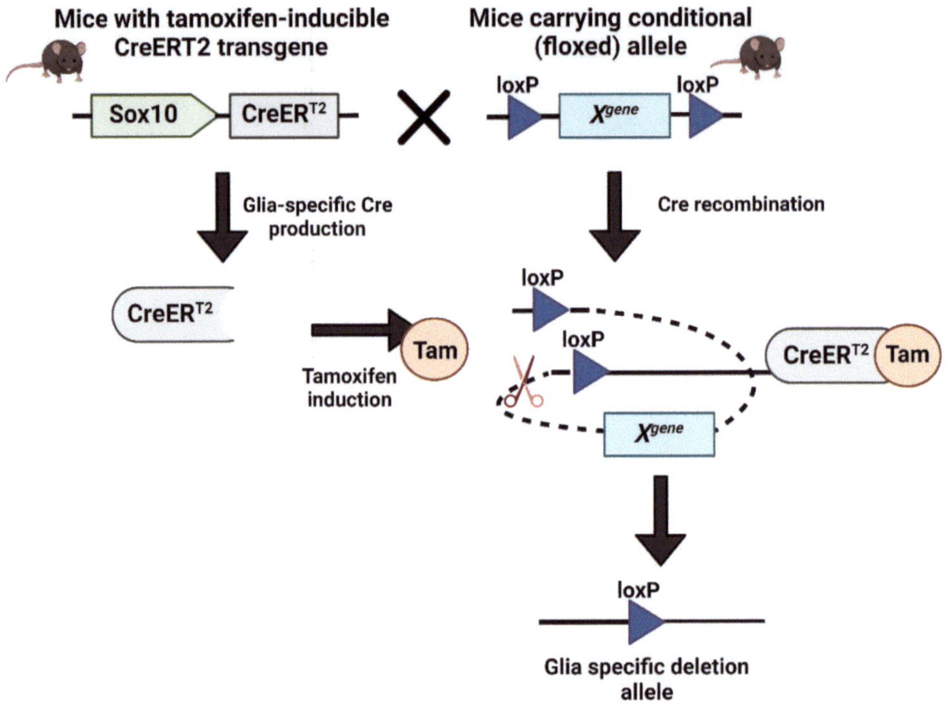

Fig. 2 Tamoxifen-inducible conditional gene targeting of enteric glia on *Sox10*^{CreERT2} mice. Schematic representation of glia-specific conditional and tamoxifen (Tam) inducible genetic deletion in *Sox10*^{CreERT2} mice. Spatially controlled glia-specific mutagenesis can be achieved by selective expression of *Sox10*^{CreERT2}. Crossing of *Sox10*^{CreERT2} mice with mice carrying a conditional or floxed allele, the floxed gene is rapidly removed in response to Tam induction. In the absence of Tam, *CreER*^{T2} is retained in the cytoplasm. Binding of Tam to the mutant ligand-binding domain of *CreER*^{T2} results in translocation of the Cre recombinase into the nucleus, where it can recombine and delete floxed DNA

DNA sequences flanked by loxP sites, resulting in the deletion of floxed sequences in Sox10 expressing cells. Sox10^{CreERT2}-driven gene deletion in enteric glia facilitates a nuanced understanding of the molecular mechanisms underlying enteric glial function in health and pathophysiology of the GI [8, 9].

2 Materials

2.1 Materials for the Selective Activation of Glia with DREADDs

1. Clozapine-N-oxide (CNO).
2. CNO 40× stock solution: Dissolve 2.4 mg of powder in 1 mL of dimethyl sulfoxide (DMSO), and it can be stored at −20 °C. Protect from light. For in vivo applications.
3. CNO 10 mM stock solution in DMSO: Stock solution can be aliquoted and stored at −20 °C for later use. Protect from light. For ex vivo studies.
4. Dimethyl sulfoxide (DMSO).

5. BD 27 G X 1/2″ hypodermic needles.
6. Sterile syringes for single use (1 mL).
7. Sterile phosphate buffer saline (PBS; 1×) or modified 1× Krebs buffer.

2.2 Materials for the Cre-Inducible Ablation of Glia Function with Tamoxifen

Tamoxifen is a hazardous substance and should be stored, handled, and disposed of following product Safety Data Sheet (SDS) recommendations and biosafety guidelines per institution.

2.2.1 Tamoxifen Diet

Chow containing tamoxifen citrate (400 mg/kg).

2.2.2 Tamoxifen IP Injection

1. Tamoxifen free base.
2. Ethanol, for molecular biology.
3. 70% ethanol, for disinfection.
4. Sunflower oil.
5. Water bath.
6. BD 27 G X 1/2″ hypodermic needles.
7. Sterile syringes for single use (1 mL).

3 Methods

3.1 Methods for the Selective Activation of Glia Using DREADDs

GFAP::hM3Dq mice can be used in vivo and ex vivo studies such as Ca^{2+} imaging [1, 6, 13] and whole gut transit [6]. Below are detailed protocols that describe approaches to stimulate glia with CNO.

1. For in vivo experiments, prepare a 60 μg/mL working solution of CNO by adding 25 μL of the 40× CNO stock to 975 μL of sterile PBS (1×) (*see* **Note 1**).
2. Using a sterile 1 mL syringe with a 27 G X 1/2″ hypodermic needle, inject mice intraperitoneally (IP) with 5 μL of working CNO solution for every 1 g of body weight (i.e., a mouse that has a body weight of 20 g will receive an injection of 100 μL). This will result in mice receiving a final dose of 0.3 mg/kg of CNO (*see* **Note 2**).
3. Effects of CNO activation after IP injection onset as early as 5 min post-injection and peaks at 30 min (*see* **Note 3**) [6, 18, 23].
4. To use for ex vivo experiments, prepare a working solution of 10 μM in modified 1× Krebs buffer or PBS and apply directly onto tissue or perfuse in tissue bath (*see* **Note 4**).

3.2 Methods for the Cre-Inducible Ablation of Glia Function with Tamoxifen

Tamoxifen induction regimens can vary for specific mouse lines and the experimental setup involved. The approaches described below have been validated for the induction of *Cre* in enteric glia from male and female adult *Sox10-Cre*^{ERT2} mice [8, 9, 11, 22].

3.2.1 Tamoxifen Administered via Diet

1. In mice with a tamoxifen-sensitive deletion of a gene specific to enteric glia (i.e., *Sox10CreERT2;Cx43f/f* [8–10] or *Sox10-CreERT2;Hnmt* [22]), *CreERT2* activity is induced by feeding animals with chow containing tamoxifen citrate (400 mg/kg) for 2 weeks.
2. After 2 weeks on a tamoxifen diet, mice should return to regular chow for at least 1 week before running experiments (*see* **Note 5**).

3.2.2 Tamoxifen Administered via IP Injection

1. Prepare a stock solution of tamoxifen at a concentration of 10 mg/mL in a 9:1 solution of sunflower oil and ethanol (*see* **Note 6**).
2. For a total volume of 10 mL, add 100 mg of tamoxifen and 1 mL of ethanol to a 15 mL conical tube.
3. Mix solution in an ultrasonic bath for 15 min at 37 °C or until tamoxifen powder is fully dissolved in ethanol.
4. Alternatively, the solution can be warmed to 37 °C in a water bath and vortexed occasionally until fully mixed.
5. Once tamoxifen powder is fully dissolved, add 9 mL of sunflower oil for a total volume of 10 mL.
6. Solution should be mixed once again in an ultrasonic bath or water bath at 37 °C until homogeneous.
7. Tamoxifen solution can then be aliquoted and stored at −20 °C for future use. Aliquots can typically be kept for up to 3 months if stored properly.
8. Using a sterile 1 mL syringe with a 27 G X 1/2″ hypodermic needle, inject mice with 10 μL of tamoxifen solution (10 mg/mL) for every 1 g of body weight (i.e., a mouse that has a body weight of 20 g will receive a 200 μL injection of tamoxifen). This will result in mice receiving a final dose of 100 mg/kg of tamoxifen.
9. Disinfect the injection site with 70% ethanol before each injection.
10. To effectively induce *CreERT2* activity, mice should receive one daily IP injection of tamoxifen free base (100 mg/kg) for 3 days. Experiments should be performed 3–5 days after the last injection (*see* **Note 7**).

11. Throughout the course of tamoxifen injections and any post-injection wait period, mice should be closely monitored for any adverse reactions to the treatment.

4 Notes

1. Chemogenetic experiments should include control groups both for the intended effect of CNO (in DREADD-expressing animals) and for a possible indirect effect of CNO itself (without DREADDs).
2. In vivo CNO doses for mice typically range from 0.25 to 5 mg/kg. Thus, the dosage can be adjusted within this range if required by the experimental paradigm. Some studies have reported off-target effects of CNO in concentrations higher than 5 mg/kg; thus, it is generally recommended to use lower doses in mice. New DREADD agonists, such as compound 21 or deschloroclozapine (DCZ), are available and reported to have the same efficacy as CNO to activate hM3Dq with relatively no off-target effect.
3. A single CNO injection of 0.3 mg/kg is sufficient to induce glial-driven physiological effects; however, serial injections of CNO can be used to study chronic effects of glia activation.
4. Ex vivo experiment CNO concentrations range from 10 to 100 μM and can be adjusted if needed [6, 13].
5. If administering tamoxifen through diet, it is important to keep track of body weight. Mice will eat less in the first few days after the diet switch, so it is essential to be mindful of animal health.
6. Tamoxifen free base solution for IP injections can also be prepared using corn oil.
7. Some recent studies have shown that tamoxifen administration can potentially impact gastrointestinal motility [24] and the microbiome [25]. Thus, it is important to use genetic controls for tamoxifen experiments to rule out any effects from the tamoxifen treatment on its own.

References

1. Seguella L, Gulbransen BD (2021) Enteric glial biology, intercellular signalling and roles in gastrointestinal disease. Nat Rev Gastroentero 18:571–587. https://doi.org/10.1038/s41575-021-00423-7
2. Boesmans W, Hao MM, Berghe PV (2018) Optogenetic and chemogenetic techniques for neurogastroenterology. Nat Rev Gastroenterol Hepatol 15:21–38. https://doi.org/10.1038/nrgastro.2017.151
3. Rao M, Nelms BD, Dong L et al (2015) Enteric glia express proteolipid protein 1 and are a transcriptionally unique population of glia in the mammalian nervous system. Glia 63: 2040–2057. https://doi.org/10.1002/glia.22876

4. Rao M, Rastelli D, Dong L et al (2017) Enteric Glia Regulate Gastrointestinal Motility but Are Not Required for Maintenance of the Epithelium in Mice. Gastroenterology 153:1068–1081.e7. https://doi.org/10.1053/j.gastro.2017.07.002
5. Woods C, Flockton AR, Wallace LE et al (2023) Proteolipid protein 1 is involved in the regulation of intestinal motility and barrier function in the mouse. Am J Physiol-Gastrointest Liver Physiol 324:G115–G130. https://doi.org/10.1152/ajpgi.00171.2022
6. McClain JL, Fried DE, Gulbransen BD (2015) Agonist-evoked Ca2+ signaling in enteric glia drives neural programs that regulate intestinal motility in mice. Cell Mol Gastroenterol Hepatol 1:631–645. https://doi.org/10.1016/j.jcmgh.2015.08.004
7. Delvalle NM, Fried DE, Rivera-Lopez G et al (2018) The cholinergic activation of enteric glia is a physiological mechanism that contributes to the regulation of gastrointestinal motility. Am J Phys Gastrointest Liver 315:G473–G483. https://doi.org/10.1152/ajpgi.00155.2018
8. Morales-Soto W, Gonzales J, Jackson WF, Gulbransen BD (2023) Enteric glia promote visceral hypersensitivity during inflammation through intercellular signaling with gut nociceptors. Sci Signal 16:eadg1668. https://doi.org/10.1126/scisignal.adg1668
9. Grubišić V, McClain JL, Fried DE et al (2020) Enteric Glia Modulate Macrophage Phenotype and Visceral Sensitivity following Inflammation. Cell Rep 32:108100. https://doi.org/10.1016/j.celrep.2020.108100
10. Grubišić V, Gulbransen BD (2017) Enteric glial activity regulates secretomotor function in the mouse colon but does not acutely affect gut permeability. J Physiol 595:3409–3424. https://doi.org/10.1113/jp273492
11. Laranjeira C, Sandgren K, Kessaris N et al (2011) Glial cells in the mouse enteric nervous system can undergo neurogenesis in response to injury. J Clin Invest 121:3412–3424. https://doi.org/10.1172/jci58200
12. Delvalle NM, Dharshika C, Morales-Soto W et al (2018) Communication between enteric neurons, glia, and nociceptors underlies the effects of tachykinins on neuroinflammation. Cell Mol Gastroenterol Hepatol 321–344: 321. https://doi.org/10.1016/j.jcmgh.2018.05.009
13. Ahmadzai MM, Seguella L, Gulbransen BD (2021) Circuit-specific enteric glia regulate intestinal motor neurocircuits. Proc Natl Acad Sci USA 118:e2025938118. https://doi.org/10.1073/pnas.2025938118
14. Brown IAM, McClain JL, Watson RE et al (2016) Enteric glia mediate neuron death in colitis through purinergic pathways that require connexin-43 and nitric oxide. Cell Mol Gastroenterol Hepatol 2:77–91. https://doi.org/10.1016/j.jcmgh.2015.08.007
15. Chow AK, Grubišić V, Gulbransen BD (2021) Enteric Glia Regulate Lymphocyte Activation via Autophagy-Mediated MHC-II Expression. Cell Mol Gastroenterology Hepatology 12:1215–1237. https://doi.org/10.1016/j.jcmgh.2021.06.008
16. Fogarty EA, Kitzman JO, Antonellis A (2020) SOX10-regulated promoter use defines isoform-specific gene expression in Schwann cells. BMC Genomics 21:549. https://doi.org/10.1186/s12864-020-06963-7
17. Lu J, Wang D, Xu J et al (2023) New Insights on the Role of Satellite Glial Cells. Stem Cell Rev Rep 19:358–367. https://doi.org/10.1007/s12015-022-10460-7
18. Xie AX, Lee JJ, McCarthy KD (2017) Ganglionic GFAP+ glial Gq-GPCR signaling enhances heart functions in vivo. JCI Insight 2:e90565. https://doi.org/10.1172/jci.insight.90565
19. Bossuyt J, Herrewegen YVD, Nestor L et al (2023) Chemogenetic modulation of astrocytes and microglia: State-of-the-art and implications in neuroscience. Glia 71:2071–2095. https://doi.org/10.1002/glia.24390
20. Agulhon C, Boyt KM, Xie AX et al (2013) Modulation of the autonomic nervous system and behaviour by acute glial cell Gq protein-coupled receptor activation in vivo. J Physiol 591:5599–5609. https://doi.org/10.1113/jphysiol.2013.261289
21. Seguella L, McClain JL, Esposito G, Gulbransen BD (2022) Functional intraregional and interregional heterogeneity between myenteric glial cells of the colon and duodenum in mice. J Neurosci. https://doi.org/10.1523/jneurosci.2379-20.2022
22. McClain JL, Morales-Soto W, Gonzales J et al (2023) Sexually dimorphic effects of histamine degradation by enteric glial histamine n-methyltransferase (HNMT) on visceral hypersensitivity. Biomolecules 13:1651. https://doi.org/10.3390/biom13111651
23. Xie AX, Madayag A, Minton SK et al (2020) Sensory satellite glial Gq-GPCR activation alleviates inflammatory pain via peripheral adenosine 1 receptor activation. Sci Rep 10:14181.

https://doi.org/10.1038/s41598-020-71073-z

24. Bhave S, Ho WLN, Cheng K et al (2022) Tamoxifen administration alters gastrointestinal motility in mice. Neurogastroenterol Motil 34:e14357. https://doi.org/10.1111/nmo.14357

25. Li H, Gao X, Chen Y et al (2023) Potential risk of tamoxifen: gut microbiota and inflammation in mice with breast cancer. Front Oncol 13:1121471. https://doi.org/10.3389/fonc.2023.1121471

Chapter 9

Enteric Glial Cell Culture to Mimic In Vivo Obesity and Neurodegenerative Disorders

Vanessa D'Antongiovanni, Clelia Di Salvo, Laura Benvenuti, and Matteo Fornai

Abstract

When exposed to inflammatory and infectious stimuli, enteric glial cells (EGCs) exhibit an abnormal phenotype (phenomenon designated as reactive gliosis), releasing a plethora of inflammatory mediators that could contribute to the onset and maintenance of inflammatory responses in the gut. In this chapter, we will describe the most useful techniques adopted to explore EGCs' response to different immune/ inflammatory stimuli, such as bacterial toxin and enteric neurodegenerative disorders (NDs)-related protein.

Key words Enteric gliosis, Enteric inflammation, Neurodegenerative disorders, Obesity

1 Introduction

In the last years, growing evidence has documented the important role of enteric glia in maintaining intestinal homeostasis [1, 2]. In particular, it has been observed that EGCs, historically considered to contribute only to the structural and trophic protection of enteric neurons, are involved in the modulation of several physiological gut functions, including motility, secretion, immune response, and mucosal barrier integrity, through a dynamic interplay with neurons, immune cells, and epithelial cells [2, 3]. Accordingly, morphological and functional alterations of EGCs have been reported to contribute to the pathophysiological mechanisms underlying digestive and extra-digestive disorders, such as inflammatory bowel diseases, obesity, and neurodegenerative disorders [1, 4, 5]. In particular, under pathological conditions characterized by a leaky gut, EGCs acquire a proinflammatory phenotype as a consequence of their interaction with pathogens or proinflammatory insults translocated into the mucosa. Such changes trigger a broad spectrum of alterations, such as activation of the

Giuseppe Esposito (ed.), *Enteric Glial Cells: Methods and Protocols*, Methods in Molecular Biology, vol. 2971,
https://doi.org/10.1007/978-1-0716-4795-0_9,

proinflammatory signaling pathways NF-κB/p38MAPK and toll-like receptors (TLRs), as well as increased expression and release of glia-related proteins and enzymes, including GFAP, S100-β, inducible nitric oxide synthase, and metalloproteinase [6, 7]. Of note, this cascade of events leads to the release of inflammatory mediators, including interleukin (IL)1-β, IL-6, and tumor necrosis factor (TNF), thus contributing to further exacerbate the intestinal epithelial barrier (IEB) damage and the neuroinflammatory response in the gut.

Metabolic disorders related to a hyperlipidemic diet intake, such as obesity, besides being associated with gut microbiota alterations [8], are characterized by an impairment of the mucous layer and IEB integrity, thus favoring the translocation of luminal noxious molecules into the mucosa [9]. The continuous influx of luminal antigens activates the mucosal immune system, including mucosal enteric glia, thus eliciting the onset of a low-grade systemic inflammation, that contribute to further exacerbate the disruption of intestinal mucosal barrier integrity and the pathological enteric neuronal remodeling processes underlying the onset of enteric motor abnormalities observed in obese patients [10, 11]. In this context, several studies observed that activated EGCs contribute to the maintenance of enteric inflammatory processes and to the onset of enteric motor abnormalities associated with obesity through the release of IL-1β and substance P (SP) [12–14].

Likewise, patients with neurodegenerative disorders, such as Alzheimer's disease (AD) and Parkinson's disease (PD), are characterized by an impairment of IEB permeability, enteric accumulation of NDs-related proteins along with signs of enteric inflammation, which, in turn, contribute to gastrointestinal disturbances complained by these patients [15–17]. Of note, clinical and pre-clinical evidence have reported a significant increase in GFAP-positive glial cells in colonic biopsies from patients and animal models of neurodegenerative disorders [18–20]. In particular, an increase in colonic glial-derived proinflammatory markers, such as IL-1β, IL-6, SP, and TNF, in colonic tissues has been observed [5, 21], indicating that activated EGCs sustain the intestinal inflammatory processes that contribute to the development of bowel motor abnormalities and mucosal barrier impairment.

Based on this interesting evidence and on the emerging role of EGCs in the pathophysiological mechanisms underlying digestive and extra-digestive diseases, this chapter provides a detailed methodology to investigate the contribution of EGCs under conditions mimicking the in vivo exposure to a hypercaloric diet and the in vivo contact with the NDs-related proteins. Such in vitro investigations provide controlled environments to better dissect specific molecular and/or cellular mechanisms, as well as allow a better investigation of cell–cell interaction than in vivo experiments.

2 Materials

2.1 Cells

Rat EGCs can be purchased from ATCC (ATCC® CRL-2690—http://www.atcc.org/). Cells are delivered in dry ice and must either be immediately processed or stored in the vapor phase of liquid nitrogen until needed for cultivation.

2.2 Reagents and Solutions

1. Dimethyl sulfoxide (DMSO). Store at room temperature.
2. Phosphate-buffered saline (PBS) 1×. Liquid. Store at +4 °C.
3. RIPA buffer: 10% Nonylphenol, ethoxylated, 8.8% sodium chloride, 6.1% Tris, 5% sodium deoxycholate, 1.12% sodium pyrophosphate decahydrate, 1% sodium lauryl sulfate, 0.38% EGTA, 0.29% EDTA, 0.22% beta glycerophosphate, 0.18% sodium orthovanadate. Store at room temperature.
4. Protease inhibitors cocktail. Store at +4 °C.
5. Trypsin–EDTA: Sterile-filtered, 2.5 g porcine trypsin and 0.2 g EDTA (0.25%). Store in aliquots at −20 °C to avoid freeze/thaw cycle.
6. 70% Ethanol solution.
7. Dulbecco's Modified Eagle's Medium (DMEM): High glucose with 10% fetal bovine serum (FBS) + 5% Penicillin/Streptomycin (P/S; 100 IU/mL) + 2 mM glutamine. Details: 500 mL of DMEM added with 50 mL of FBS, 5 mL of P/S, and 5 mL of glutamine. Do not freeze.
8. Palmitate (PA)-bovine serum albumin (BSA) complex: 200 mM of PA stock solution must be dissolved in pre-heated water. The PA-BSA complex is prepared, prior to use, by mixing the PA stock solution with 10% BSA solution (BSA powder + culture medium) (*see* **Note 1**).
9. 100 μg/mL of lipopolysaccharide (LPS) stock solution: Solubilize 1 mg of LPS powder in 10 mL of sterile PBS. Mix the solution until all the powder is solubilized and store at −80 °C.
10. 221 μM Amyloid-β (Aβ) solution: Solubilize 1 mg of Aβ powder in 75 μL of NH_4OH plus 925 μL of PBS. Mix the solution until all the powder is solubilized and store at −80 °C.
11. 50 μM α-synuclein (α-syn) solution: Solubilize 500 mg of α-syn powder in 500 μL of distilled water. Mix the solution until all the powder is solubilized and store at −80 °C.

3 Methods

The present protocols have been developed on the immortalized culture of EGCs. All cell manipulation procedures are conducted under a laminar flow cabinet, respecting good laboratory practice to work safely.

3.1 Frozen Cells Handling and Culture Maintenance

1. Thaw the vial containing the frozen cells by gentle agitation in a 37 °C water bath. Thawing should be rapid, approximately 2 min.
2. Remove the vial from the water bath as soon as the contents are thawed and decontaminate by spraying with 70% ethanol.
3. Transfer the vial contents to a 15 mL collection tube with an appropriate volume of pre-warmed complete culture medium (we recommend 5 mL of medium).
4. To remove the cryoprotective agent, centrifuge the cell suspension at 1200 rpm for 5 min at room temperature.
5. Discard the supernatant and resuspend the pellet with 5 mL of pre-warmed complete culture medium.
6. Transfer the cell suspension into a 75 cm^2 tissue culture flask, containing 5–6 mL of pre-warmed complete culture medium.
7. Incubate the culture at 37 °C in a suitable incubator with 5% CO_2 in an air atmosphere.

3.1.1 Subculturing

1. View cultures using an inverted microscope to assess the degree of confluence and proceed to cell splitting at close to 80–90% confluence.
2. Remove the medium and rinse with 5 mL of PBS twice.
3. Aspirate the PBS and add 1 mL of trypsin–EDTA solution.
4. Incubate the flask at 37 °C until the cells detach.
5. Add 5 mL of pre-warmed complete culture medium.
6. Centrifuge the cell suspension at 1200 rpm for 5 min at room temperature.
7. Discard the supernatant and resuspend the pellet with pre-warmed complete culture medium to obtain a ratio of 1:4 to 1:6 in new culture flasks. These flasks should then be topped up with culture media to the required volume.

3.2 Experimental Procedure to Mimic the In Vivo Exposure to a Hypercaloric Diet Intake

In order to investigate the role of enteric glia in the pathophysiological mechanisms underlying the intestinal inflammation associated with obesity, EGCs are treated with PA (the major source of fat in the hypercaloric diet) and LPS (reflect the endotoxemia associated with high-fat diet (HFD) intake) [13, 14]. At the end of the experimental procedure, cultured media are collected to assess the inflammatory mediators released from EGCs (i.e., IL-1β, IL-6, and SP) while cells are lysed to evaluate the activation of EGCs (i.e., GFAP and S100-β) and the expression of proteins involved in inflammatory signaling pathways (i.e., TLRs, inflammasome components) (Fig. 1a and Table 1).

1. Seed the cells: 1×10^6 cells in Petri dishes (p100) or 1×10^5 in a 6-well plate in pre-warmed complete culture medium.

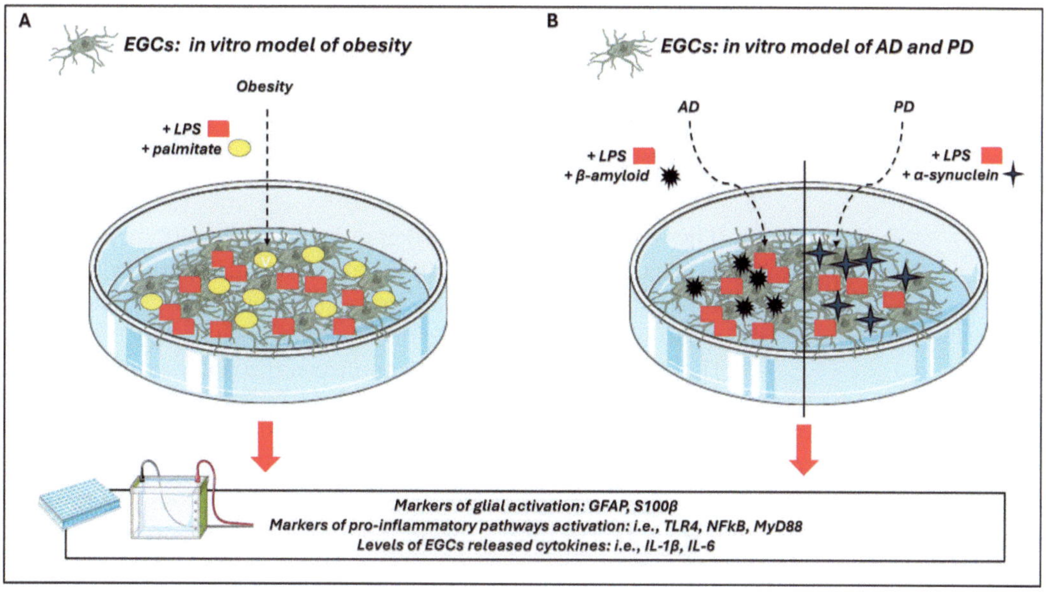

Fig. 1 Schematic representation of in vitro models of EGCs mimicking obesity (**a**) and neurodegenerative disorders (AD and PD) (**b**). To mimic the in vivo exposure to HFD, EGCs are treated with 400 μM PA and 10 μg/mL LPS (**a**). To mimic the in vivo features of AD and PD, EGCs are treated with 5 μg/mL of LPS and 1 μM Aβ or 1 μM α-syn, respectively (**b**). At the end of the experiment, the cells can be lysed and the supernatants collected for subsequent analyses. *Abbreviations*: Aβ: amyloid-β, α-syn: α-synuclein, AD: Alzheimer's disease, HFD: high-fat diet, LPS: lipopolysaccharide, PA: palmitate, PD: Parkinson's disease

Table 1
List of the primary antibodies and ELISA Kits for cell lysates and supernatants derived from EGCs

Protein	Primary antibody
Western blot	
Caspase-1	ab1872, Abcam
S100 calcium binding protein B (S100-β)	ab52642, Abcam, Cambridge, United Kingdom
Glial fibrillary acidic protein (GFAP)	ab53554, Abcam
Nuclear factor-κB p65 (NF-κB p65)	sc-8008, Santa Cruz, Dallas, United States
Phospho-JNK	#4668S, Cell Signaling
Toll-like receptor 4 (TLR4)	ab22048, Abcam
Kit ELISA	
Interleukin (IL)-1β	ab100767, Abcam
Substance P (SP)	ab288318, Abcam

2. After 24 h, cells were treated with PA (400 μM) for 72 h. Control cells were run in parallel. The same final concentration of BSA (0.5% v/v) was maintained in all treated cells to avoid a differential protein binding effect on compounds.
3. Incubate cells in a humidified incubator at 37 °C with 5% CO_2.
4. After 3 days, add LPS (10 μg/mL).
5. The next day, cells are lysed and the supernatants are collected. In particular, supernatants should be centrifuged for 5 min at 1200 rpm to obtain a cell-free medium. Then, stock supernatants at −80 °C to perform further analysis (i.e., release of IL-1β from EGCs). For what concern the cell lysis protocol, *see* Subheading 3.5.

3.3 Experimental Procedure to Mimic the In Vivo Neurodegenerative Disorders

Patients with neurodegenerative disorders, such as PD and AD, display signs of intestinal inflammation along with intestinal NDs-related protein accumulation and enteric gliosis [19, 22]. Accordingly, to investigate the role of enteric glia in this pathological setting, EGCs are treated with LPS (to reflect an altered intestinal permeability) and Aβ or α-syn, to mimic AD and PD, respectively [21]. At the end of the experimental procedure, cultured media are collected to assess pro-inflammatory mediators release from EGCs, while cells are lysed to evaluate the activation of EGCs (i.e., GFAP and S100-β) and the expression of proteins involved in inflammatory signaling pathways (i.e., TLRs, inflammasome components) (Fig. 1b and Table 1).

1. Seed the cells: 1×10^6 cells in Petri dishes (p100) or 1×10^5 in a 6-well plate in pre-warmed complete culture medium.
2. After 24 h, cells were treated with LPS (10 μg/mL) for 4 h.
3. After 4 h, add Aβ 1 μM (to mimic AD) or α-syn 1 μM (to mimic PD) and incubate cells for 72 h. Control cells were run in parallel.
4. Incubate cells in a humidified incubator at 37 °C with 5% CO_2.
5. After 3 days, cells are lysed and the supernatants are collected. In particular, supernatants should be centrifuged for 5 min at 1200 rpm to obtain a cell-free medium. Then, stock supernatants at −80 °C to perform further analysis (i.e., release of IL-1β from EGCs). For what concern the cell lysis protocol, *see* Subheading 3.5.

3.4 Conditioned Medium Derived from Activated EGCs to Study the Interaction Between Enteric Glia and Other Intestinal Cell Subtypes

1. Medium derived from cultured EGCs (designed as "Conditioned medium"), treated with PA and LPS (to mimic the in vivo exposure to HFD) or with Aβ or α-syn (to mimic the in vivo features of neurodegenerative disorders) should be centrifuged for 5 min at 1200 rpm to obtain a cell-free medium.

2. Next, conditioned media should be used to treat other intestinal cell subtypes, such as macrophages, intestinal epithelial [23] cells, and enteric neurons, as well as in contact with intestinal organoids. The time of incubation should be optimized based on the cell type used and on the effects that we want to study.
3. Of note, in order to better investigate the contribution of enteric glia in IEB alterations, conditioned media should also be used to treat intestinal organoids in 3D culture in order to increase in vitro adherence with in vivo conditions. Also, in this case, the time of incubation should be optimized based on the effects that we want to study.

3.5 Cell Lysis

1. Wash cells directly in the tissue culture plate by adding cold PBS. Aspirate PBS and repeat one time. Keep the tissue culture dish on ice throughout.
2. Add 80 μL/well of RIPA buffer 1× plus protease inhibitors cocktail (*see* **Note 2**). Gently scrape the entire surface of each well using a cell scraper until most of the cells are removed (*see* **Note 3**).
3. Transfer in a microfuge tube and grind the samples on ice using about 100 strokes (*see* **Note 4**). Then, sonicate on ice each sample for 5–6 s with 60% of ultrasound amplification (repeat this point three times).
4. Vortex the samples every 5 min for a total of 30 min (*see* **Note 4**).
5. Centrifuge the microfuge tubes at 12,000 rpm for 30 min at 4 °C. Transfer supernatant to a fresh tube and store at −80 °C to perform further analysis (i.e., evaluation of glial activation markers).

4 Notes

1. To minimize variability, the same final concentration of BSA (0.5% v/v) should be maintained also in untreated cells (control cells).
2. Protease inhibitor cocktail is formulated as a pill. Add 1 pill each 10 mL of RIPA buffer 1×.
3. To confirm cells total removed, check the surface of each well at the microscope.
4. After each mixing, put the tube on ice to avoid sample overheating.

References

1. D'Antongiovanni V, Pellegrini C, Antonioli L et al (2023) Enteric glia and brain astroglia: complex communication in health and disease along the gut-brain axis. Neuroscientist 30: 493. https://doi.org/10.1177/10738584231163460
2. Gulbransen BD, Christofi FL (2018) Are we close to targeting enteric glia in gastrointestinal diseases and motility disorders? Gastroenterology 155:245–251. https://doi.org/10.1053/J.GASTRO.2018.06.050
3. Sharkey KA (2015) Emerging roles for enteric glia in gastrointestinal disorders. J Clin Invest 125:918–925. https://doi.org/10.1172/JCI76303
4. Seguella L, Gulbransen BD (2021) Enteric glial biology, intercellular signalling and roles in gastrointestinal disease. Nat Rev Gastroenterol Hepatol 18:571–587. https://doi.org/10.1038/S41575-021-00423-7
5. Benvenuti L, D'antongiovanni V, Pellegrini C et al (2020) Enteric glia at the crossroads between intestinal immune system and epithelial barrier: implications for Parkinson disease. Int J Mol Sci 21:1–16. https://doi.org/10.3390/IJMS21239199
6. Cirillo C, Sarnelli G, Turco F et al (2011) Proinflammatory stimuli activates human-derived enteroglial cells and induces autocrine nitric oxide production. Neurogastroenterol Motil 23:e372. https://doi.org/10.1111/J.1365-2982.2011.01748.X
7. Kabouridis PS, Lasrado R, McCallum S et al (2015) The gut microbiota keeps enteric glial cells on the move; prospective roles of the gut epithelium and immune system. Gut Microbes 6:398–403. https://doi.org/10.1080/19490976.2015.1109767
8. Panattoni A, Calvigioni M, Benvenuti L et al (2022) The administration of Enterococcus faecium SF68 counteracts compositional shifts in the gut microbiota of diet-induced obese mice. Front Microbiol 13. https://doi.org/10.3389/FMICB.2022.1054097
9. D'antongiovanni V, Segnani C, Ippolito C et al (2023) Pathological remodeling of the gut barrier as a prodromal event of high-fat diet-induced obesity. Lab Investig 103:100194. https://doi.org/10.1016/J.LABINV.2023.100194
10. Camilleri M, Malhi H, Acosta A (2017) Gastrointestinal complications of obesity. Gastroenterology 152:1656–1670. https://doi.org/10.1053/J.GASTRO.2016.12.052
11. D'Antongiovanni V, Pellegrini C, Fornai M et al (2020) Intestinal epithelial barrier and neuromuscular compartment in health and disease. World J Gastroenterol 26:1564–1579. https://doi.org/10.3748/WJG.V26.I14.1564
12. Seguella L, Pesce M, Capuano R et al (2021) High-fat diet impairs duodenal barrier function and elicits glia-dependent changes along the gut-brain axis that are required for anxiogenic and depressive-like behaviors. J Neuroinflammation 18:115. https://doi.org/10.1186/S12974-021-02164-5
13. D'Antongiovanni V, Benvenuti L, Fornai M et al (2020) Glial A2B adenosine receptors modulate abnormal Tachykininergic responses and prevent enteric inflammation associated with high fat diet-induced obesity. Cells 9. https://doi.org/10.3390/cells9051245
14. Antonioli L, D'Antongiovanni V, Pellegrini C et al (2020) Colonic dysmotility associated with high-fat diet-induced obesity: role of enteric glia. FASEB J 34:5512–5524. https://doi.org/10.1096/FJ.201901844R
15. Fu P, Gao M, Yung KKL (2020) Association of Intestinal Disorders with Parkinson's Disease and Alzheimer's Disease: a systematic review and meta-analysis. ACS Chem Neurosci 11: 395–405. https://doi.org/10.1021/ACSCHEMNEURO.9B00607
16. Wang F, Fei M, Hu WZ et al (2022) Prevalence of constipation in elderly and its association with dementia and mild cognitive impairment: a cross-sectional study. Front Neurosci 15. https://doi.org/10.3389/FNINS.2021.821654
17. Puig KL, Lutz BM, Urquhart SA et al (2015) Overexpression of mutant amyloid-β protein precursor and presenilin 1 modulates enteric nervous system. J Alzheimer's Dis 44:1263–1278. https://doi.org/10.3233/JAD-142259
18. Pellegrini C, D'antongiovanni V, Ippolito C et al (2021) From the intestinal mucosal barrier to the enteric neuromuscular compartment: an integrated overview on the morphological changes in Parkinson's disease. Eur J Histochem 65. https://doi.org/10.4081/EJH.2021.3278
19. Bellini G, Benvenuti L, Ippolito C et al (2023) Intestinal histomorphological and molecular alterations in patients with Parkinson's disease. Eur J Neurol 30:3440–3450. https://doi.org/10.1111/ENE.15607

20. Clairembault T, Kamphuis W, Leclair-Visonneau L et al (2014) Enteric GFAP expression and phosphorylation in Parkinson's disease. J Neurochem 130:805–815. https://doi.org/10.1111/JNC.12742
21. D'Antongiovanni V, Pellegrini C, Antonioli L et al (2021) Palmitoylethanolamide counteracts enteric inflammation and bowel motor dysfunctions in a mouse model of Alzheimer's disease. Front Pharmacol 12. https://doi.org/10.3389/FPHAR.2021.748021
22. Li Y, Chen Y, Jiang L et al (2021) Intestinal inflammation and Parkinson's disease. Aging Dis 12:2052. https://doi.org/10.14336/AD.2021.0418
23. D'Antongiovanni V, Fornai M, Colucci R et al (2024) Enteric glial NLRP3 inflammasome contributes to gut mucosal barrier alterations in a mouse model of diet-induced obesity. Acta Physiol 17:e14232. https://doi.org/10.1111/apha.14232

Chapter 10

Fluorescence-Activated Cell Sorting of Primary Enteric Glia

Austin Irwin, Anthony Pellicano, and Vladimir Grubišić

Abstract

Enteric glia are important regulators of the gastrointestinal homeostasis in health and disease. To better understand these processes, it is necessary to be able to isolate and study enteric glia. Here we provide a protocol that expands upon conventional methods of glial cell isolation with the addition of microfluidic cell sorting, exploiting the glia-specific expression of the ACSA2 surface protein. Isolated live cells can be used for a number of downstream processes, such as cell culturing and/or proteomic or genomic analysis. Thus, the detailed isolation methods used are potentially effective ways of isolating and enriching enteric glia.

Key words Enteric nervous system (ENS), Enteric glial cells, Fluorescent activated cell sorting, Flow cytometry, ACSA2

1 Introduction

A complex interplay between enteric glia and surrounding cells is required to uphold proper functioning of the gastrointestinal system [1], yet these glia are poorly characterized and understood [2]. Despite this deficit, enteric glia have been found to have wide-ranging roles such as the regulation of gut motility, the secretomotor reflex, epithelial barrier function, and visceral sensitivity [3]. Research into enteric glia can provide further insight into the progression and manifestation of many gut disorders, such as irritable bowel syndrome, infections, inflammatory bowel disease, and cancer [4]. For this research to progress, it is critical to develop more advanced methods in the isolation and manipulation of enteric glia.

The ideal method of enteric glia isolation must be one that is simple and inexpensive while retaining a high degree of specificity. The complex network of cell types and tissue connections that glia reside in makes it difficult to isolate only one specific cell type without the appropriate tools. Many methods of isolation rely on

Giuseppe Esposito (ed.), *Enteric Glial Cells: Methods and Protocols*, Methods in Molecular Biology, vol. 2971,
https://doi.org/10.1007/978-1-0716-4795-0_10,

the use of specific growth factors in cell culture, but this approach also changes the phenotype of the glia [5] and, hence, does not allow for investigation of the enteric glia in situ.

Here, we expand upon established methods of enteric glia isolation [6]. Conventional methods to enrich glia cannot be used on live cells because they make use of unexposed intracellular proteins, such as s100b, GFAP, and Sox10. Recent studies have explored the use of the glia-specific cell surface proteins, such as Astrocyte Cell Surface Marker 2 (ACSA2), to allow for live enteric glia isolation [7]. We propose that through the use of these new surface markers, such as ACSA2, and modern cell sorters capable of sorting the cells by distinct fluorescent signals, we can label and subsequently sort enteric glia by exploiting the use of labeled antibodies targeting ACSA2. Following the isolation protocol, the purified product can be used for downstream culture or various omics profiling.

2 Materials

2.1 Ethical Statement

IACUC Protocol Committee.

2.2 Animals

C57BL/6 J, RRID: IMSR_JAX:000664.

2.3 Sorting Antibodies and Viability Dye

1. FcR Blocking Reagent, mouse.
2. Anti-ACSA-2-PE, Clone IH3-18A3.
3. Calcein-AM.

2.4 Generic Lab Equipment

1. Pipette and pipette tips (covering a range of 1–1000 μL).
2. Dissection microscope.
3. Cell-culture hood.
4. Refrigerated 15 mL/50 mL centrifuge.
5. Refrigerated microcentrifuge.
6. Dissection tools.
7. 1.5 mL microcentrifuge tubes.
8. Sterile cell-culture-grade 1.5 mL microcentrifuge tubes.
9. 50 mL conical tubes.
10. 15 mL conical tubes.
11. 40 μm filter.
12. Cotton swabs.
13. 10 mL syringe.
14. Pipettes and pipette tips.

15. 37 °C water bath or bead bath.
16. Liquid containers (to hold media).
17. Source of ultrapure water (reverse osmosis water dispenser, deionized water dispenser + autoclave, individual packages, etc.).
18. pH meter.
19. Sterile cell strainers: 40 μm.
20. Sterile filtration method (either/or):
    (a) Syringe filter: Basix Syringe Filters, nylon, sterile.
    (b) 50 mL conical pump filter: Steriflip sterile disposable vacuum filter units.
    (c) (Recommended) 150/500 mL pump filter: Bottle-top vacuum filter system.
21. Cell-culture pump/vacuum (depending on sterile filtration method used).

2.5 Cell-Sorting Equipment

1. Pala Single and Bulk Cell Sorter.
2. Bulk Sorting Cartridge.

2.6 Media and Reagents

1. Non-sterile DMEM (1 L): 1 Packet of DMEM, 1 L of ultrapure H_2O, sodium bicarbonate—1.2 g.
2. Sterile complete medium: ~447.5 mL of non-sterile DMEM, 2.5 mL of L-glutamine–penicillin–streptomycin solution—50 units/mL, FBS (10%)—50 mL, *Sterile Filtration* (*see* Subheading 2.4, **item 20**).
3. Digestion buffer (enough for four colons, 2.5 mL per colon): 1 mL of HBSS (10×), calcium, magnesium, no phenol red, 23.8 mg of HEPES (10 mM final concentration), 100 μL of deoxyribonuclease I from bovine pancreas (100 Kunitz/mL final concentration), 100 μL of liberase (0.13 Wünsch units/mL final concentration), 8.8 mL of ultrapure water (milli-Q water system).
4. 10× phosphate-buffered saline (PBS): Sodium phosphate dibasic, Na_2HPO_4—10.9 g, sodium phosphate monobasic, NaH_2PO_4—3.2 g, sodium chloride, NaCl—90 g, ultrapure H_2O—bring up to 1 L, check/adjust pH to 7.2.
5. 1× PBS: 10× PBS (Subheading 2.6, **item 4**)—100 mL, ultrapure H_2O—bring up to 1 L.
6. 1× PBS with 0.1% triton X-100: 1 L 1× PBS, 1 mL of Triton X-100.
7. 10× PBS with EDTA (1 L): Sodium phosphate dibasic, Na_2HPO_4—10.9 g, sodium phosphate monobasic, NaH_2PO_4—3.2 g, sodium chloride, NaCl—90 g, solid or

liquid EDTA: (solid EDTA (20 mM)—5.845 g; (recommended) liquid EDTA (0.5 M)—40 mL, ultrapure H_2O—bring up to 1 L, check/adjust pH to 7.2.

8. 1× PBS with EDTA (1 L): 10× PBS with EDTA—100 mL, ultrapure H_2O—bring up to 1 L.
9. 1× PBS with EDTA and DNAse (50 mL): 1× PBS with EDTA—50 mL, deoxyribonuclease I from bovine pancreas—100 μL, *Sterile Filtration* (*see* Subheading 2.4, **item 20**).
10. 1% BSA solution: 1× PBS (Subheading 2.6, **item 5**)—10 mL, bovine serum albumin (IgG-free, protease-free; BSA)—0.100 g.

3 Methods

3.1 Tissue Dissection and Dissociation (Modified Protocol from [6])

1. Euthanize mouse with method approved by ethics committee (cervical dislocation, CO_2 asphyxiation, rapid decapitation).
2. Spray the surface of the mouse with 70% ethanol.
3. Surgically remove the large intestine, spanning from the ileocecal valve to the anus.
4. Immediately place a colon in cooled DMEM medium solution (Subheading 2.6, **item 1**, *see* **Note 1**).
5. Gently flush the colon with cold DMEM (Subheading 2.6, **item 1**) using a 10 mL syringe fitted with a medium pipette tip to remove feces and intestinal contents. Repeat until the colon is clear of obstruction.
6. Remove any remaining mesentery and connective tissue with forceps and surgical dissection scissors.
7. Cut colons into individual 2–3 cm sections for easier manipulation.
8. Insert a plastic or glass rod into the colon and ensure there is high tension within the tissue without ripping. Darker colored rods will offer the best contrast.
9. While observing under a dissection microscope, press cotton swabs onto the surface of the colon and with minimal pressure, gently tease away the outer longitudinal muscle myenteric plexus layer (LMMP; containing the myenteric plexus and circular muscle layers) from the interior regions.
10. Continue this pattern from the top colon segment to the bottom, gradually turning the colon around on the rod to remove other areas.
11. After most of the LMMP has been removed, use a combination of cotton swabs, forceps, and scissors to pull and cut the "teased apart" LMMP fully from the rest of the tissue still on the rod.

12. Transfer isolated LMMP into cold non-sterile DMEM (Subheading 2.6, **item 1**) on ice and continue dissection of the other colons previously cut into 2–3 cm sections (in Subheading 3.1, **step 8**).
13. Repeat Subheading 3.1, **steps 1 → 12** for each mouse. Larger and older mice will have larger colons and require more time for complete dissection compared to the colons of smaller and younger mice. Older and (potentially) more sick/fibrotic colons will have more connective tissue than younger and more healthy colons and thus may be harder to dissect.
14. When all LMMP sections have been isolated, transfer sections to a plate containing 2.5 mL of regular PBS or digestion buffer to wash off excess non-sterile DMEM.
15. Hold the LMMP sections with forceps over a gentleMACS TM C-tube containing 2.5 mL of preheated digestion solution (Subheading 2.6, **item 2**) and cut the sections into smaller pieces using fine scissors. Colon pieces may get stuck on the scissors, in which case you can simply dip the scissors into the digestion solution or deposit the tissue on the side of the tube.
16. If needed, tubes are either flung with an aggressive downwards motion or tissue is physically moved via surgical forceps to ensure tissue sections are not stuck to the top portions of the tube but are properly resting in the digestion solution on the bottom.
17. Incubate for 30 min at 37 °C in a water or bead bath, gently mixing the tube by hand every 10 min (*see* **Note 2**). Incubation times longer than 30 min can result in large-scale die-off of enteric glia.
18. After 30 min digestion, dissociate the digested tissue using a gentleMACS™ Dissociator on program: [m_intestine01]**.
19. To terminate the digestion reaction, add 10 mL of *complete* medium (Subheading 2.6, **item 3**) with FBS.
20. Proceed with filtering the slurry: Place a 40 μm strainer on top of an open 50 mL conical tube and add some complete medium to fully pre-wet the filter prior to adding the cell slurry. After pre-wetting, slowly add the cell slurry to the filter. After all the slurry has been used, wash the filter briefly with some complete medium again. There will be large, undigested/undissociated chunks present on the filter; this is normal.
21. Transfer the filtered mixture to a 15 mL tube.
22. Centrifuge at 300 rcf/G for 10 min at 4 °C.
23. Discard the supernatant so that only 1–2 mL remains. Resuspend the pellet in the remaining supernatant by pipetting up and down. Ensure no bubbles during the resuspension process (*see* **Note 3**).

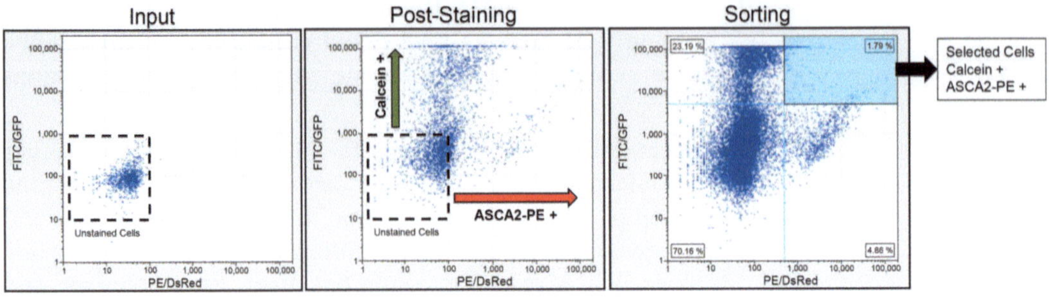

Fig. 1 Flow cytometry and FACS of LMMP-digested single cell suspension. Unstained cells (*Input*) were simultaneously stained with ASCA2-PE antibody (measured along 561 nm Excitation and PE/DsRed Emission axis) for enteric glia and stained with Calcein (measured along 488 nm excitation and FITC/GFP emission axis) to select for living cells (*Post-Staining*). Cells with robust staining for both ASCA2-PE antibody and Calcein labeling were identified as the enteric glia population (*Sorting*)

3.2 PALA Cell Sorter

1. Prior to immunolabeling, get a reading of background fluorescence by loading a small sample of the resuspended pellet from Subheading 3.1, **step 23** into the flow cytometer/cell sorter. Save and record values and use as a comparison to evaluate the success of staining.
2. Add 10 μL of Miltenyi FcR Blocking Reagent per 10^7 cells and incubate cells for 15 min at 4 °C (*see* **Note 4**).
3. Add 2–4 μL of anti-ACSA2 antibody and 2 μL of Calcein-AM and incubate cells for 15 min at 4 °C (*see* **Notes 4 and 5**).
4. Wash cells by adding 1–2 mL of buffer per 10^7 cells and centrifuge at 300 *g* for 10 min at 4 °C.
5. Aspirate supernatant completely and resuspend pellet in 1× PBS with EDTA. We suggest at least 1 mL of buffer, but different cell sorters will have different requirements.
6. Proceed with fluorescence-activated cell sorting (FACS) of the immunolabeled cells according to the cell sorter's instructions. Compare the results to the results of the background in (Subheading 3.2, **step 1**) to determine if staining of live enteric glia was successful (*see* Fig. 1 for an example).
7. Use sorted cells for downstream applications.

4 Notes

1. As the colon is an unsterile organ, initial procedures will be done with minimally sterile methods, with increasing sterility closer to the final cell sorting.
2. While the dissected tissue is in the digestion buffer. Periodically agitate the digestion mixture every 5–10 min to ensure the maximal surface area of the tissue is exposed.

3. Prevention of bubbles in the media is paramount. Bubbles may interfere with both the culturing process and the isolation protocol.
4. Do not vortex the FcR Blocking reagent or the anti-ACSA2 antibody.
5. The time of incubation for the anti-ACSA2 antibodies is time-sensitive and generally should not greatly exceed 15 min. However, should a concentration of greater than 10^7 cells be used, then increased incubation times may be required.

References

1. Rao M, Gulbransen BD (2024) Enteric Glia. Cold Spring Harb Perspect Biol 17:a041368. https://doi.org/10.1101/cshperspect.a041368
2. Rosenberg HJ, Rao M (2021) Enteric glia in homeostasis and disease: from fundamental biology to human pathology. iScience 24(8):102863. https://doi.org/10.1016/j.isci.2021.102863
3. Seguella L, Gulbransen BD (2021) Enteric glial biology, intercellular signalling and roles in gastrointestinal disease. Nat Rev Gastroenterol Hepatol 18(8):571–587. https://doi.org/10.1038/s41575-021-00423-7
4. Santhosh S, Zanoletti L, Stamp LA, Hao MH, Matteoli G (2024) From diversity to disease: unravelling the role of enteric glial cells. Front Immunol 15:1408744. https://doi.org/10.3389/fimmu.2024.1408744
5. Jessen KNR, Saffrey MJ, Burnstock G (1983) The enteric nervous system in tissue culture. I. Cell types and their interactions in explants of the myenteric and submucous plexuses from Guinea pig, rabbit and rat. Brain Res 262:17–35. https://doi.org/10.1016/0006-8993(83)90466-3
6. Smith TH, Ngwainmbi J, Grider JR, Dewey WL, Akbarali HI (2013) An in-vitro preparation of isolated enteric neurons and glia from the myenteric plexus of the adult mouse. J Vis Exp 78:50688. https://doi.org/10.3791/50688
7. Grubišić V, Gulbransen BD (2022) Astrocyte cell surface antigen 2 and other potential cell surface markers of enteric glia in the mouse colon. ASN Neuro 14:17590914221083203. https://doi.org/10.1177/17590914221083203

Chapter 11

Methods to Isolate Enteric Glia from hPSC-Derived Enteric Ganglioids

Evgeniia Pimenova, Faranak Fattahi, and Homa Majd

Abstract

Enteric glia play pivotal roles in maintaining the enteric nervous system (ENS) homeostasis, supporting neuronal functions, and initiating and developing enteric neuropathies. However, understanding their diversity and functional roles has been formidably challenging due to difficulties in isolating these cells from human tissue. Here, we describe a detailed method for generating enteric glia from human pluripotent stem cells (hPSCs). This approach involves differentiating hPSCs into enteric glial cells through a vagal and enteric neural crest intermediates, replicating in vivo developmental processes. The hPSC-derived enteric glia express canonical markers (SOX10, GFAP, PLP1, S100B) and display the cellular diversity of primary human enteric glia, confirmed by single-nucleus RNA sequencing (snRNA-seq). These cultures are scalable, enabling high-throughput screening and molecular analysis, and they facilitate the study of glial interactions with other cell types, including immune cells, epithelial cells, and the gut microbiome. hPSC-derived enteric glia represent a significant advancement in enteric neurobiology, providing a versatile platform for disease modeling and therapeutic discovery, thereby improving our understanding of enteric glial cell functions in both health and disease.

Key words Peripheral glia, DGBI, In vitro differentiation, Drug screening, Enteric neurons, Gut, Gliosis

1 Introduction

Comprising a sophisticated network of millions of neurons and glia, the enteric nervous system (ENS) extends throughout the entire gastrointestinal (GI) tract and controls various gut activities, including motility, secretions, absorption, and local blood flow. Enteric glia are crucial for ENS physiology and disease mechanisms, contributing to ENS homeostasis, neuronal support, intestinal barrier function, GI reflexes, neuroinflammatory cascades, progenitor generation, and the release of neurotrophic factors, as well as interaction with various gut cells, including immune and epithelial cells [1–6].

Giuseppe Esposito (ed.), *Enteric Glial Cells: Methods and Protocols*, Methods in Molecular Biology, vol. 2971,
https://doi.org/10.1007/978-1-0716-4795-0_11,

Emerging evidence suggests that enteric glia play dynamic roles in a range of diseases, including neuroinflammation, cancer, and infections, often adopting pro-inflammatory or pro-tumorigenic phenotypes in these conditions [7–10]. Understanding the precise mechanisms by which enteric glia influence GI physiology and how these roles are altered in disease states remains a major area of research. This highlights the need for further exploration to unravel the complexities of their functional repertoire. Such insights hold promise for advancing our comprehension of ENS function and may have implications for therapeutic interventions targeting gastrointestinal disorders [8, 11–14].

Despite this, persistent technical challenges in accessing and manipulating these cells within gut tissue have left them significantly understudied. Accessing human ENS cells requires invasive biopsies or surgeries, as they make up only 1% of gut tissue and are deeply embedded within muscular and mucosal layers [15]. Furthermore, the large-scale isolation and purification of ENS cells remain highly challenging. The lack of specific markers for enteric glial cells poses a significant challenge in targeting them in vivo. Common markers such as SOX10, GFAP, S100B, and PLP1 are also found in other glial cells within the peripheral and central nervous systems. This prevents the use of genetic labeling techniques like Cre/lox recombination to selectively target enteric glia. Most research relies on manipulating single markers, which may not represent the entire population of enteric glia, leaving many aspects of their biology unexplored. There are no well-established surface markers for FACS-based purification of enteric glia and their subtypes [13, 15–20].

Animal models fail to fully replicate human ENS (patho)-physiology, as rodents can tolerate mutations that cause severe enteric neuropathies in humans. Mouse models are not well-suited for large-scale and systematic studies like high-throughput screening or omics research, particularly when targeting rare or difficult-to-isolate cell types. These models are also associated with long experimental timelines and high costs. Moreover, the murine ENS differs from the human ENS in glia-to-neuron ratios, neuronal subtype balance, and the melanocortin system. Hence, it is currently impractical to isolate and culture enteric glial cells in sufficient quantities for high-throughput genetic and pharmacological experiments. Isolating ENS cells ex vivo poses additional challenges, as enteric glial cells are scarce in the GI tract, requiring large tissue samples from many animals. Standard dissection and digestion techniques are harsh, leading to low cell yield and viability [15].

Culturing human enteric glial cells in vitro offers unique opportunities to study their intrinsic properties, secretome, and responses to various perturbations, leading to important discoveries about their roles in inflammation and interactions with gut cells.

However, obtaining these cells is difficult, relying on fresh tissue from GI surgeries often affected by disease, which may affect experimental outcomes. The harsh extraction methods reduce cell yield and diversity, while variability in tissue sources and gut regions further complicates studies. The inherent diversity of enteric glia, differing by gut region and layer, adds to the variability and limits the throughput of these cultures.

We propose hPSC-derived in vitro ENS cultures as alternative models that address many of these challenges, facilitating significant advances in enteric neurobiology. hPSCs present a promising in vitro source for generating enteric neurons and glia, providing valuable models for studying human ENS development and pathology, as well as potential cell therapy applications. This chapter details a novel differentiation strategy to derive enteric glia from hPSCs through a vagal neural crest intermediate.

Recent advancements have created 2D hPSC-derived ENS cell cultures and 3D enteric ganglioids, mimicking primary ENS developmental programs by sequentially inducing vagal and enteric neural crest cells and differentiating them into ENS progenitors and mature neurons, with glial cells emerging over time, which is consistent with the in vivo developmental timeline [21–24]. These hPSC-based models capture the cellular diversity of the human ENS, as assessed by marker expression and unbiased snRNA-seq transcriptomics. All together, these data indicate that our hPSC-derived enteric glia recapitulate the transcriptional, functional, and regional features of the primary human enteric glia. They are scalable and suitable for high-throughput screens and detailed molecular profiling. They express canonical markers (SOX10, GFAP, PLP1, S100B) and include diverse subtypes similar to primary human enteric glia, providing insights into the functional differences between glial subtypes. Furthermore, using a stem cell differentiation protocol, patient-derived induced pluripotent stem cells (iPSCs) can be differentiated into enteric glial cells to investigate their role in the pathogenesis of various GI disorders.

These instances illustrate how an hPSC differentiation system can enhance our understanding of enteric glial biology and its functions in both normal and diseased states, complementing existing research models.

2 Materials

The cell lines we used are human embryonic stem cell (hESC) line H9 (WA-09, WiC, RRID: CVCL_1240), hESC line UCSF4 (University of California, San Francisco, RRID:CVCL_A022), and hiPSC line WTC11 (Coriell Institute, Gladstone Institute, RRID: CVCL_Y803).

2.1 Medium Preparation (see Notes 1 and 2)

1. hPSC maintenance medium: Mix Essential 8-Flex supplement (20 μL mL^{-1}) with Essential 8 Flex medium.
2. Medium A: Add BMP4 (1 ng mL^{-1}), SB431542 (10 μM), and CHIR 99021 (600 nM) to Essential 6 medium.
3. Medium B: Add SB431542 (10 μM), and CHIR 99021 (1.5 μM) to Essential 6 medium.
4. Medium C: Add SB431542 (10 μM), CHIR 99021 (1.5 μM), and RA (1 μM) to Essential 6 medium.
5. Neural crest cell (NCC) medium: Add FGF2 (10 ng mL^{-1}), CHIR 99021 (3 μM), N2 supplement (10 μL mL^{-1}), B27 supplement (20 μL mL^{-1}), glutagro (10 μL mL^{-1}), and MEM NEAAs (10 μL mL^{-1}) to neurobasal medium.
6. Enteric neural crest (ENC) medium: Add GDNF (10 ng mL^{-1}), ascorbic acid (100 μM), N2 supplement (10 μL mL^{-1}), B27 supplement (20 μL mL^{-1}), glutagro (10 μL mL^{-1}), and MEM NEAAs (10 μL mL^{-1}) to neurobasal medium.
7. SCM: Schwann cell medium, supplement (no FBS), normocin (1:1000).
8. Ethylenediaminetetraacetic acid (EDTA) 1×: Dilute 500 mM EDTA (1000×) to a final concentration of 500 μM in phosphate buffered saline (PBS) (Ca2+- and Mg2+-free).

2.2 All Solutions

1. Dulbecco's Modified Eagle Medium/Nutrient Mixture F-12, DMEM/F-12: (no glutamine).
2. Accutase.
3. Retinoic acid/RA.
4. SB431542.
5. Geltrex™ LDEV-Free.
6. Fibronectin, human/FN.
7. Laminin I, mouse/LM.
8. Poly-L-ornithine hydrobromide/PO.
9. 0.4% (wt/vol) trypan blue solution.
10. Normocin.

3 Methods

3.1 Coating of Culture Plates

1. Geltrex-coated plates: Quickly thaw and dilute a 500 μL frozen aliquot of geltrex in 50 mL of chilled DMEM/F12 by vigorously pipetting using a 50 mL serological pipette. Coat the wells with the diluted geltrex solution (100 μL per cm^2 of well surface area) and incubate them overnight at 37 °C. Aspirate the coating medium before use. To prevent gelatinization of

geltrex, perform this step as quickly as possible and use cold DMEM/F12 for dissolving the frozen aliquot (*see* **Note 3**).

2. PO/FN/laminin-coated plates: Prepare a 15 μg mL^{-1} poly-ornithine (PO) solution in PBS by diluting a 1000 x stock. Coat wells with 100 μL of PO/PBS solution per cm^2 of well surface area and incubate the plates overnight at 37 °C. The following day, aspirate PO/PBS and coat the wells with 100 μL per cm^2 of well surface area of a freshly made FN/laminin/PBS solution containing 2 μg mL^{-1} of fibronectin (FN) and 2 μg mL^{-1} of laminin. Plates can be used after a minimum of 2 h incubation at 37 °C. Aspirate FN/laminin/PBS before use (*see* **Note 4**).

3.2 Maintenance of hPSC Culture

1. Aspirate hPSC medium from the hPSC culture and add fresh medium (2 mL/well 6-well plate) every other day until colonies are ~80% confluent.
2. To passage, aspirate hPSC medium and wash cells with 1 mL/well 6-well plate (*see* **Note 5**).
3. Add 500 μM EDTA (2 mL/well 6-well plate). Replace the lid of the plate. Watch through an inverted microscope and monitor the cells for detachment of colony edges (4–5 min) (*see* **Note 6**).
4. Use a 5- or 10 mL serological pipette to transfer the same volume of hPSC medium into each well and mechanically harvest the cells by pipetting up and down a few times.
5. Transfer the cell suspension into a 15 mL conical tube.
6. Spin the cells (2 min, 290× *g*, 20–25 °C) and carefully discard the supernatant.
7. Resuspend the cells in an appropriate volume of hPSC medium and transfer to wells of a new gelatin-coated plate (*see* **Note 7**).
8. Incubate at 5% CO_2 and 37 °C (*see* **Note 8**).

3.3 Enteric Neural Crest Induction (Figs. 1, 2, 3 and 4)

1. On day 0, replace hPSC maintenance medium with Medium A (200 μL of medium per cm^2 of well surface area) (*see* **Notes 9** and **10**).
2. On day 2, cells should be 100% confluent. Gently aspirate spent Medium A and feed cells with Medium B (200 μL of medium per cm^2 of well surface area).
3. On day 4, feed cells with Medium B (200 μL of medium per cm^2 of well surface area).
4. On day 6, replace Medium B with Medium C containing 10 mM SB431542, 1.5 mM CHIR99021, and 1 mM retinoic (400 μL of medium per cm^2 of well surface area).
5. On days 8 and 10, feed as day 6.

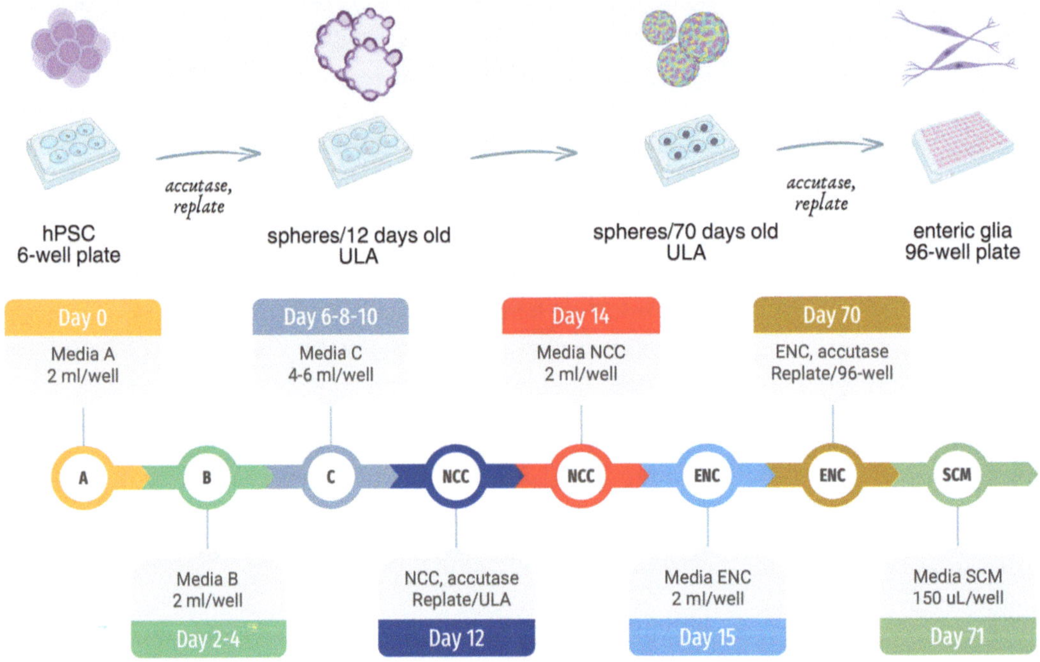

Fig. 1 Human pluripotent stem cells are employed in a three-dimensional culture system to recapitulate this differentiation process. Abbreviations: hPCS human pluripotent stem cells, ULA ultra-low attachment plate, NCC neural crest cells medium, ENC enteric neural crest medium

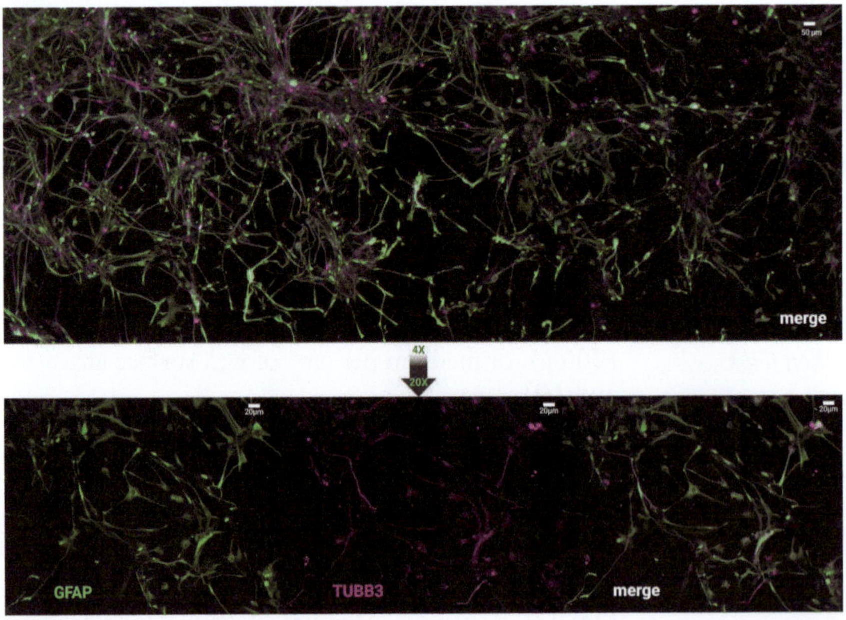

Fig. 2 Immunofluorescent microscopy images of cells obtained according to the presented protocol (Keyence, BZ-X series, 4× and 20×)

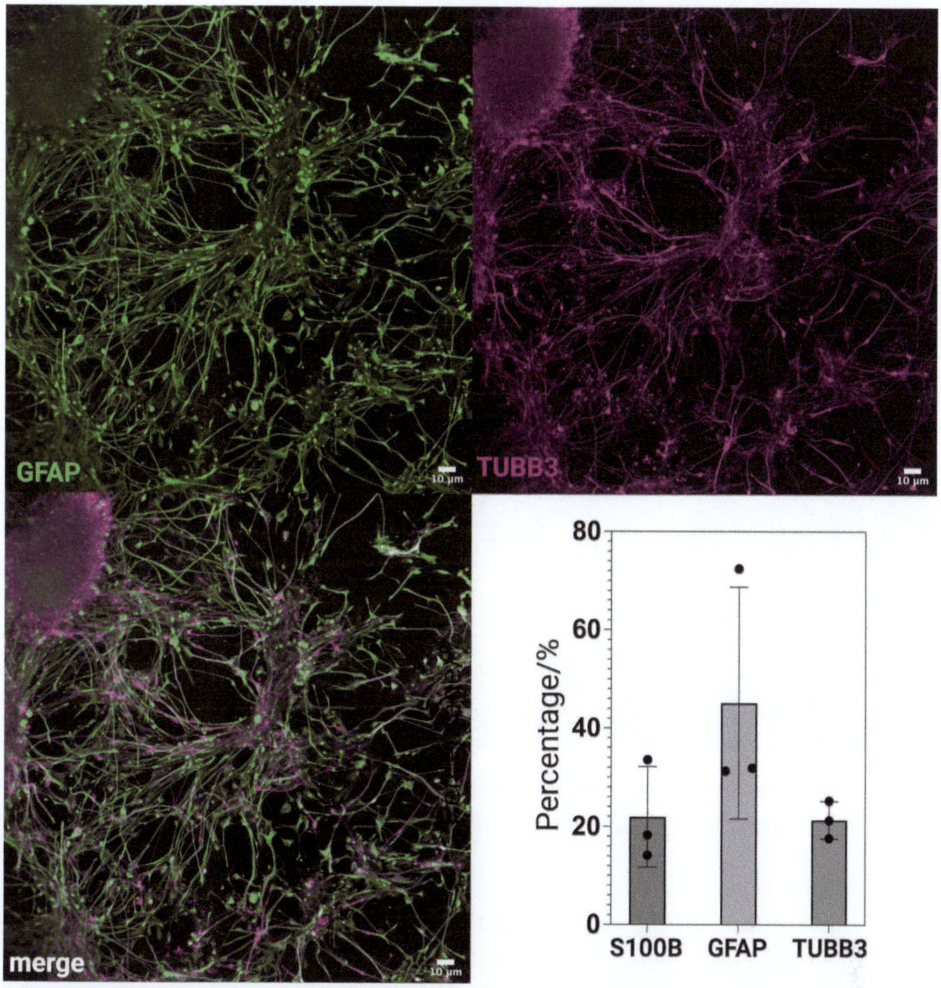

Fig. 3 Immunofluorescent microscopy images of cells obtained according to the presented protocol (ImageXpress Micro Confocal, 10×) and flow cytometry of cells obtained according to the presented protocol (Attune; four lasers: 405 nm, 488 nm, 561 nm, 637 nm; 14 fluorescence channels)

6. On day 12, gently aspirate Medium C and wash cells with 1 mL/well/6-well plate. Add the same volume of Accutase and incubate for 30 min at 5% CO_2 and 37 °C (*see* **Note 11**).
7. Dilute Accutase by adding the same volume of NCC medium. Using a serological pipette, harvest the single cell suspension and add it to a 15 mL conical tube.
8. Spin the cells (2 min, 290× *g*, 20–25 °C) and carefully discard the supernatant.
9. Using a 10 mL serological pipette, add 12 mL (for a full 6-well plate) of NCC medium and resuspend the cells completely by mechanically pipetting up and down 1–2 times. Transfer the cell suspension to an ultra-low attachment plate (ULA).

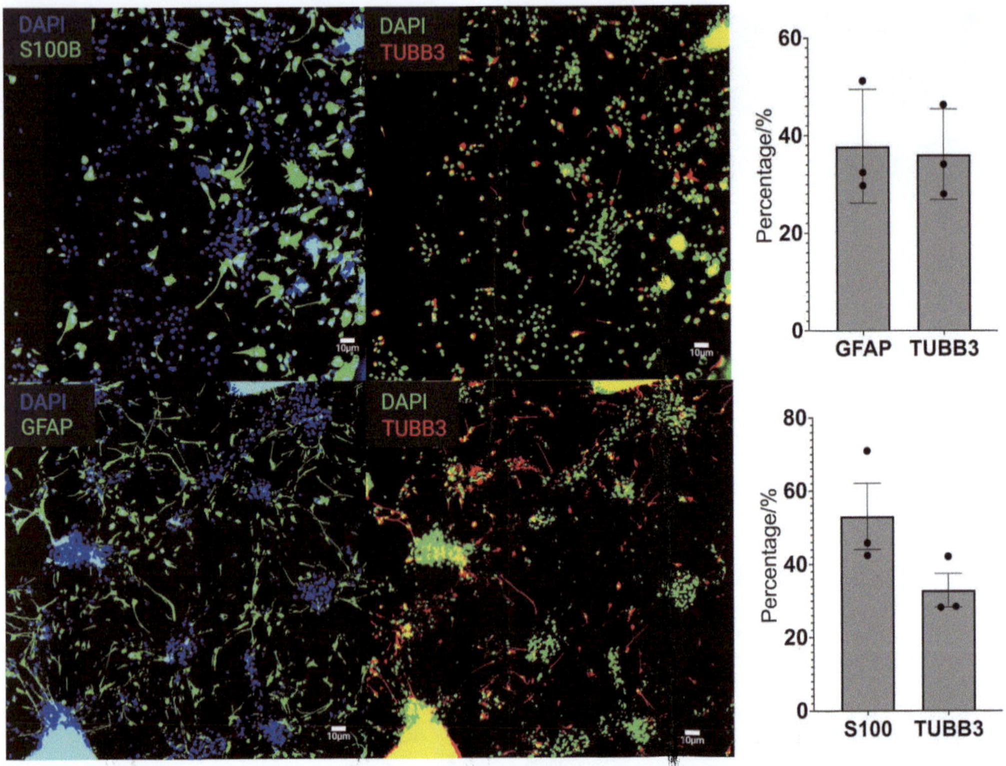

Fig. 4 Immunofluorescent microscopy images of cells obtained according to the presented protocol, masks (ImageXpress Micro Confocal, 10×, software Fiji). The percentage of cells stained with GFAP, S100, and TUBB3

10. On day 14, gently swirl to get all the cells in the center. Then, use a P1000 to aspirate the medium, moving the pipette in a circle, so cells are not aspirated.
11. On day 15, using the same swirling motion, remove NCC medium and feed cells with 2 mL/well ENC medium.
12. Feed cells with 2 mL/well ENC medium every other day until day 40.
13. Feed cells with 3 mL/well ENC medium supplemented with antibiotic (normocin 1:1000) twice a week until day 60.
14. Feed cells with 3–4 mL/well ENC medium supplemented with antibiotic (normocin 1:1000) once a week until day 80.
15. On day 80, gently swirl to gather all the cells in the center. Remove the medium. Add Accutase (100 μL per cm^2 of well surface area) and dissociate the spheroids by incubating for 60–90 min at 5% CO_2 and 37 °C. Note. Spheroids are dense on day 80.

16. Add DMEM/F-12 1 mL/well, resuspend the spheroids, and transfer them to the 15 mL conical sterile tube and centrifuge at 290× *g* for 2 min.
17. Aspirate supernatant and resuspend the pellet in 5 mL of ENC medium.
18. Transfer the cells with multichannel pipettes to the PO/FN/laminin-coated 96-well plate (at 10,000–20,000 cells per well).
19. Feed glial cells with 150–200 μL/well Schwann Cell Medium (SCM medium+supplement + antibiotic (normocin 1:1000)) every day.

4 Notes

1. Concentrations mentioned throughout the protocol are final concentrations of the medium components.
2. Prepare all medium under sterile conditions in a laminar flow hood, and store at 4 °C in the dark. Use within 2 weeks.
3. Geltrex temperature should be kept cold during thawing (on ice) and aliquoting (tubes on ice) to prevent coagulation.
4. Stock aliquots should be stored at −80 °C. Once thawed, each aliquot should be kept at 4 °C to avoid multiple freeze/thaw cycles and used within 4 weeks.
5. Use Ca^{2+} and Mg^{2+}-free PBS throughout this protocol unless otherwise stated.
6. Optimal EDTA incubation time for passaging depends on the hPSC line. Avoid vigorous pipetting for detaching colonies, as it can lead to excessive colony dissociation and cell death. When passaging to start a differentiation plate, incubate cells with EDTA for 1–2 min longer.
7. For regular hPSC maintenance, use a 1:18–1:24 passaging ratio (1:18 means transferring one-third of the cells dissociated from a single well of a 6-well plate into a full new 6-well plate). To start a differentiation plate, follow a 5:6 passaging ratio.
8. All cell incubation steps are at 5% CO_2 and 37 °C in a humidified incubator.
9. For optimal NC induction, start medium A treatment (day 0) when cells are about 80% confluent. This will be achieved within 1–2 days when cells are passaged at a 5:6 ratio.
10. As cultures grow during the NC induction, cells may detach from the underlying monolayer. Gently remove old medium and add fresh medium to the side of the well to avoid excess cell loss during feeding.

11. On day 12, approximately a full 6-well NC induction plate is transferred into a full 6-well ultra-low attachment plate after Accutase treatment (1:1 ratio).

References

1. Coelho-Aguiar Jde M, Bon-Frauches et al (2015) The enteric glia: identity and functions. Glia 63(6):921–935. https://doi.org/10.1002/glia.22795
2. Seguella L, Gulbransen BD (2021) Enteric glial biology, intercellular signalling and roles in gastrointestinal disease. Nat Rev Gastroenterol Hepatol 18:571–587. https://doi.org/10.1038/s41575-021-00423-7.3
3. Wang Z, Ocadiz-Ruiz R, Sundaresan S et al (2018) Isolation of enteric glial cells from the Submucosa and Lamina Propria of the adult mouse. J Vis Exp (138):57629. https://doi.org/10.3791/57629
4. Grubišić V, Gulbransen BD (2017) Enteric glia: the most alimentary of all glia. J Physiol 595(2):557–570. https://doi.org/10.1113/JP271021
5. Bali V, Grubišić V (2024) Enteric glia as friends and foes of the intestinal epithelial barrier function. Front Immunol 30(15):1394654. https://doi.org/10.3389/fimmu.2024.1394654
6. Scavuzzo MA, Letai KC, Maeno-Hikici Y et al (2023) Enteric glial hub cells coordinate intestinal motility. BioRxiv Prepr Serv Biol. 2023.06.07.544052. https://doi.org/10.1101/2023.06.07.544052
7. Findlay Q, Yap KK, Bergner AJ et al (2014) Enteric neural progenitors are more efficient than brain-derived progenitors at generating neurons in the colon. Am J Physiol Gastrointest Liver Physiol 307(7):G741–G748. https://doi.org/10.1152/ajpgi.00225.2014. Epub 2014 Aug 14
8. Ochoa-Cortes F, Turco F, Linan-Rico A et al (2016) A new frontier in neurogastroenterology and clinical target for inflammatory bowel diseases. Inflamm Bowel Dis 22(2):433–449. https://doi.org/10.1097/MIB.0000000000000667
9. Santhosh S, Zanoletti L, Stamp LA et al (2024) From diversity to disease: unravelling the role of enteric glial cells. Front Immunol 18(15):1408744. https://doi.org/10.3389/fimmu.2024.1408744
10. Le Berre C, Naveilhan P, Rolli-Derkinderen M (2023) Enteric glia at center stage of inflammatory bowel disease. Neurosci Lett 13(809):137315. https://doi.org/10.1016/j.neulet.2023.137315
11. Rosenberg HJ, Rao M (2021) Enteric glia in homeostasis and disease: from fundamental biology to human pathology. iScience 24(8):102863. https://doi.org/10.1016/j.isci.2021.102863
12. Scantlen MD, Majd H, Fattahi F (2023) Modeling enteric glia development, physiology and disease using human pluripotent stem cells. Neurosci Lett 811:137334. https://doi.org/10.1016/j.neulet.2023.137334
13. Baghdadi MB, Ayyaz A, Coquenlorge S et al (2022) Enteric glial cell heterogeneity regulates intestinal stem cell niches. Cell Stem Cell 29(1):86–100.e6. https://doi.org/10.1016/j.stem.2021.10.004
14. Bubeck M, Becker C, Patankar JV (2023) Guardians of the gut: influence of the enteric nervous system on the intestinal epithelial barrier. Front Med (Lausanne) 25(10):1228938. https://doi.org/10.3389/fmed.2023.1228938
15. Drokhlyansky E, Smillie CS, Van Wittenberghe N et al (2020) The human and mouse enteric nervous system at single-cell resolution. Cell 182(6):1606–1622.e23. https://doi.org/10.1016/j.cell.2020.08.003
16. Rao M, Nelms BD, Dong L et al (2015) Enteric glia express proteolipid protein 1 and are a transcriptionally unique population of glia in the mammalian nervous system. Glia 63(11):2040–2057. https://doi.org/10.1002/glia.22876
17. May-Zhang AA, Tycksen E, Southard-Smith AN et al (2021) Combinatorial transcriptional profiling of mouse and human enteric neurons identifies shared and disparate subtypes in situ. Gastroenterology 160(3):755–770.e26. https://doi.org/10.1053/j.gastro.2020.09.032
18. Morarach K, Mikhailova A, Knoflach V et al (2021) Diversification of molecularly defined myenteric neuron classes revealed by single-cell RNA sequencing. Nat Neurosci 24(1):34–46. https://doi.org/10.1038/s41593-020-00736-x
19. Elmentaite R, Kumasaka N, Roberts K et al (2021) Cells of the human intestinal tract

mapped across space and time. Nature 597(7875):250–255. https://doi.org/10.1038/s41586-021-03852-1

20. Guyer RA, Mueller JL, Goldstein AM (2022) Applications of single-cell sequencing technology to the enteric nervous system. Biomolecules 12(3):452. https://doi.org/10.3390/biom12030452
21. Majd H, Samuel RM, Ramirez JT et al (2022) hPSC-derived enteric ganglioids model human ENS development and function. bioRxiv 2022.01.04.474746. https://doi.org/10.1101/2022.01.04.474746
22. Barber K, Studer L, Fattahi F (2019) Derivation of enteric neuron lineages from human pluripotent stem cells. Nat Protoc 14(4):1261–1279. https://doi.org/10.1038/s41596-019-0141-y
23. Fattahi F, Steinbeck JA, Kriks S et al (2016) Deriving human ENS lineages for cell therapy and drug discovery in Hirschsprung disease. Nature 531(7592):105–109. https://doi.org/10.1038/nature16951
24. Majd H, Richter MN, Samuel RM et al (2024) Differentiation of enteric nervous system lineages from human pluripotent stem cells. J Vis Exp 207. https://doi.org/10.3791/66133

Chapter 12

Calcium Imaging Studies in Enteric Glia

Beatriz Thomasi, Julia Jamka, and Brian D. Gulbransen

Abstract

Calcium (Ca^{2+}) imaging is a versatile technique to study cellular activity and is a particularly important tool to study electrically passive cells such as glia. Enteric glia play important roles in gut physiology, and their activity can be captured using Ca^{2+} imaging. Ca^{2+}-encoded activities command glial processes such as gliotransmission and intercellular signaling. This chapter describes a protocol that will allow users to study glial Ca^{2+} activity in ex vivo whole-mounts of circular muscle-myenteric plexus. This chapter will include comprehensive steps to execute Ca^{2+} imaging experiments, data extraction, and analysis.

Key words Calcium imaging, Enteric glia, Enteric nervous system, Neuro-glial interactions, Neurocircuits, Glial cells, Genetically encoded calcium indicators, GCaMP

1 Introduction

Calcium ions (Ca^{2+}) are ubiquitous intracellular signals that regulate multiple aspects of cell survival and function, including responses to external cues, transcription, metabolism, proliferation, cell motility, and intercellular signaling. As a result, intracellular Ca^{2+} ($[Ca^{2+}]_i$) is under tight control by mechanisms involving channels, pumps, and buffers [1]. Changes in $[Ca^{2+}]_i$ are of particular interest in neural systems where fluxes in $[Ca^{2+}]_i$ are tied to neuronal and glial processes underlying excitability, neurotransmission, gliotransmission, and synaptic signaling [2, 3]. Therefore, imaging $[Ca^{2+}]_i$ has become a core technique neurobiologists use to study cellular activity in neural networks in the brain and periphery. This is particularly important for studies of glial cells, which exhibit passive electrical currents and instead encode activity in the form of variations in $[Ca^{2+}]_i$.

Supplementary Information The online version contains supplementary material available at https://doi.org/10.1007/978-1-0716-4795-0_12.

Giuseppe Esposito (ed.), *Enteric Glial Cells: Methods and Protocols*, Methods in Molecular Biology, vol. 2971, https://doi.org/10.1007/978-1-0716-4795-0_12,

Early studies of glial activity in hippocampal explant cultures used organic Ca^{2+} indicator dyes to demonstrate that glial cells exhibit spontaneous Ca^{2+} events that propagate in patterns throughout cells and eventually form Ca^{2+} waves that spread through glial networks [4, 5]. These and subsequent observations formed the base knowledge that glia are active signaling nodes in neural networks that exhibit activity encoded by Ca^{2+}-based signaling mechanisms despite lacking electrical activity similar to neurons. It is now known that glial Ca^{2+} signaling is complex and that distinct subcellular Ca^{2+} responses, intrinsic fluctuations, waves, and microdomains play roles in normal physiology regulated by cells such as astrocytes within neuronal circuits. Therefore, studying Ca^{2+} activity in glia has deepened the understanding of their roles in neural signaling, regulation of neurotransmission, neuroplasticity, neurodegeneration, and behavior [6–8].

Enteric glia are perhaps the largest population of peripheral glia and are housed with enteric neurons in the walls of the intestine. Enteric glia are active signaling cells within enteric neurocircuits that regulate enteric neurotransmission and homeostasis. Similar to astrocytes in the brain, enteric glia exhibit activity encoded by variations in $[Ca^{2+}]_i$ [9–12]. Enteric glial Ca^{2+} responses are triggered by numerous external and internal cues such as neurotransmitters, neuromodulators, immune signals, and mechanical force [13, 14]. Mechanisms downstream of glial Ca^{2+} events are involved in key gastrointestinal functions such as motility, secretion, and immune defense. Therefore, recording $[Ca^{2+}]_i$ in enteric glia has become a core technique to study activity among these cells in the gut.

Several techniques are available to study $[Ca^{2+}]_i$ in enteric glia. These include organic, chemical-based dyes and genetically encoded Ca^{2+} sensors. Organic Ca^{2+} indicator dyes are small molecules that are loaded into cells and exhibit changes in fluorescence upon binding to Ca^{2+} [15]. Such dyes are typically used for short-duration experiments because dyes are eventually extruded or degraded over time. Genetically encoded calcium indicators (GECIs) are sensor proteins expressed by cells with fluorescent properties in the presence of Ca^{2+} [16, 17]. GECIs are currently the most widely used method to study Ca^{2+} responses in neural networks and have broad applicability in vitro, in vivo, and ex vivo [18]. The basic GECI system is designed based on a permuted GFP with the Ca-binding protein calmodulin (GCaMP). GECIs' expression can be targeted to specific cells and/or subcellular domains, and multiple improvements in this system have generated sophisticated GCaMPs for high-speed imaging, increased signal-to-noise ratio, and dynamic range [18–20].

This chapter provides a detailed protocol for Ca^{2+} imaging in enteric glia in live whole-mount preparations of the myenteric plexus using GECIs. Following this protocol will allow investigators to study glial responses evoked by numerous modes of activation in cellular to network level resolution.

2 Materials

2.1 Animals

For WntCre2; GCaMP5g-tdT mouse generation, commercially available mouse lines were acquired and crossed in-house. Transgenic mice expressing the genetically encoded Ca2+ indicator GCaMP5g under the transcriptional control of the Wnt1 promoter (Wnt1Cre2+/−;PC::G5-tdT+/−; hereafter, referred to as WntCre2; GCaMP5g-tdT) were generated by crossing Wnt1Cre2 mice (129S4.Cg-E2f1Tg(Wnt1-cre)2Sor/J (https://www.jax.org/strain/022137; The Jackson Laboratory; RRID: IMSR_JAX:022137) with PC::G5-tdTomato (PC::G5-tdT) mice (catalog #02447, The Jackson Laboratory; B6;129S6-Polr2aTn (pb-CAG-GCaMP5g,-tdTomato)Tvrd/J; RRID: IMSR_JAX:024477). Overall, the WntCre2; GCaMP5g-tdT animal model allows Ca^{2+} activity to be monitored in enteric glia and neurons (Wnt1+ cells). The tdT reporter is highly expressed in enteric glia, enabling high-precision detection of enteric glia in the processing and quantification steps.

2.2 Tissue Extraction and Dissection

1. Modified Krebs solution: 121 mM NaCl, 5.9 mM KCl, 21.2 mM $NaCO_3$, 1.2 mM $MgCl_2$, 2.5 mM $CaCl_2$, 1.2 mM $NaHPO_4$, 8 mM D-glucose, 10 mM HEPES, and 1 mM sodium pyruvate.
2. Fine surgical tools: Tweezers numbers 5, 7, and 5/45, scissors, micro scissors.
3. Dissection station (C-DS Diascopic Stand S).
4. Sylgard-coated Petri dishes.
5. Insect pins.

2.3 Tissue Incubation

Incubator.

2.4 Ca^{2+} Imaging Setup

1. Wide-field water-immersion objective lens (Olympus XLUMPLFLN20xW, 1.0 numerical aperture) on an upright Olympus BX51WI fixed-stage microscope (Center Valley).
2. Lumencor solid-state LED light source.
3. Photometrics Prime BSI camera.
4. Perfusion system.
5. Automatic temperature controller connected to perfusion pencil system (Warner instruments).

6. Sylgard-coated live imaging engineered dishes.
7. Peri-Star Pro peristaltic pump.
8. 1 mL syringe with valve.
9. Micromanipulator: ROI200, MPC200 (Sutter instruments).
10. Glass micropipette.

2.5 Acquisition

Metamorph.

2.6 Analysis

Fiji software.

3 Methods

3.1 Tissue Extraction and Mucosal Dissection

1. After euthanasia, open the lower abdominal cavity and expose the gastrointestinal tract with the aid of surgical scissors (*see* **Note 1**).
2. "Unroll" the small intestine with the help of forceps. Cecum identification makes it possible to reach the colon and the ileum. For ileum preparation, cut the intestine distal to the stomach and proximal to the cecum. Preparation of the large intestine involves cutting the colon above the cecum and below the rectum.
3. Immediately transfer tissue samples to a container of ice-cold Krebs solution and keep on ice. Sylgard-coated dishes are used as a flat, flexible base for fine dissection. Fill the Sylgard dissecting dish with Krebs solution and then transfer the intact piece of intestine (~3–6 cm). The following steps must be performed under a stereomicroscope (C-DS Diascopic Stand S, Nikon).
4. Stretch and secure the intestinal end with insect pins with the mesenteric border facing the operator—this way, this morphological mark will guide the opening of the gut (*see* **Note 2**).
5. Use micro scissors to make a straight cut along the mesenteric border from the distal to the proximal end of the intestinal tube.
6. Remove the feces and hold the tissue on the Sylgard-dish using insect pins. The expected orientation is shown in Fig. 1a, with the mucosal layer facing upwards.
7. Pin the tissue flat by creating a uniformly applied vector force stretching the tissue with the same tension and orientation (Fig. 1b).
8. In the most proximal region, start removing the mucosal layer using fine forceps (#5 and #5/45) (Fig. 1d). It is recommended to leave a thin layer of intestinal mucosa on the laterals of the preparation, as this helps to identify smooth muscle orientation (Fig. 1e–f).

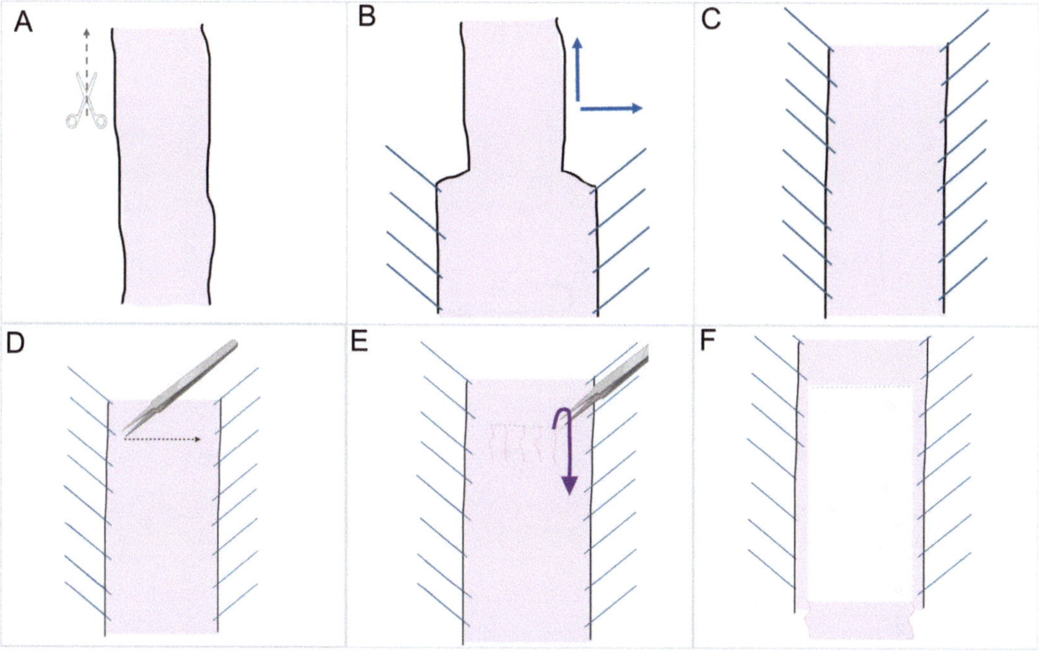

Fig. 1 Schematic illustration of the mucosal dissection for whole-mount circular muscle-myenteric plexus preparations. (**a**) Upward exposure of the mucosal layer after cutting the mesenteric border. In (**b**), the organization of the insect's pins for stretching the intestine. Note: the pins are placed in parallel pairs. Grab the edge of the intestine, and pull it slightly outward and upward, like the arrows representing vector forces. This pattern stretches the intestine with well-distributed tension as seen in (**c**). To remove the mucosa, define a starting point as the dotted line in (**d**) and remove pieces of mucosa along this defined line. Finally, dissect gently and very carefully up to the most distal portion. It is expected that at the end of this process, a considerable area of the mucosa will have been dissected, exposing the submucosal plexus

3.2 Smooth Longitudinal Muscle Microdissection

1. Cut small proportional squares of mucosa-free colon samples (~0.5 cm^2) and transfer them to the live imaging dish filled with ice-cold Krebs solution (*see* **Note 3**) (Fig. 2a).
2. Flip the tissue orientation to reach the smooth longitudinal muscle (Fig. 2b).
3. Start pinning at the four corners with mini-insect pins and then gradually stretch and pin the preparation using the same vector technique as for dissecting the mucosa (Fig. 2c).
4. Identify muscle layer orientation and use fine forceps (#5) with a delicate pinching and pulling movement to remove the longitudinal smooth muscle bundles (*see* **Note 4**) (Fig. 2c).
5. Return the live imaging dish containing circular muscle-myenteric plexus preparation (CMMP) to ice until tissue incubation or any tissue treatment is performed. After dissection, you will have a sample containing the following layers from top to bottom: myenteric plexus, circular muscle, and submucosal plexus.

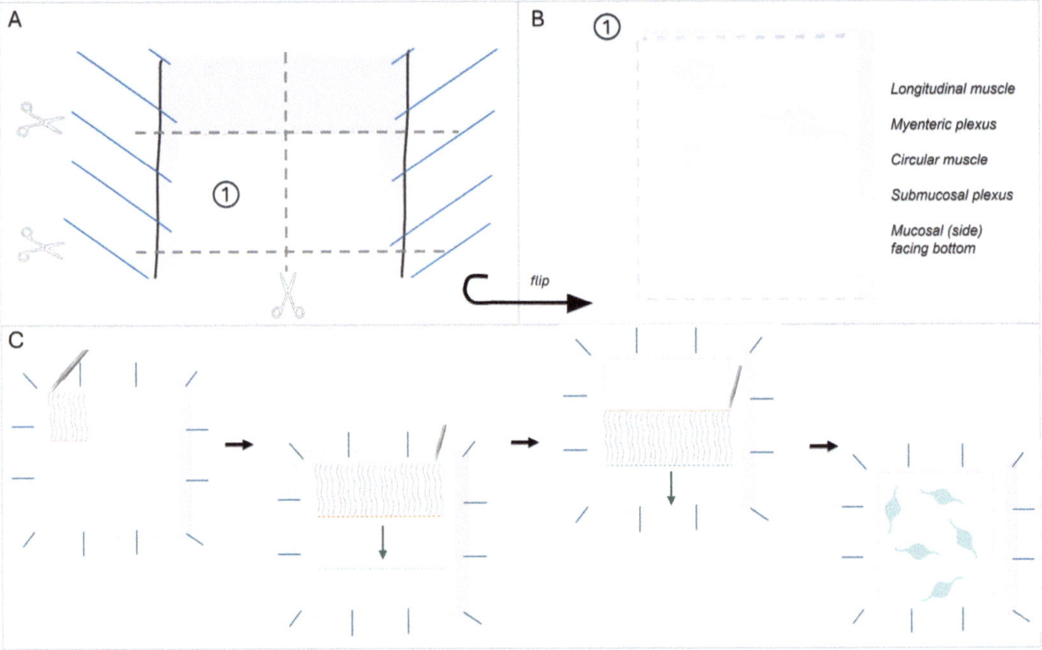

Fig. 2 Longitudinal muscle dissection. (**a**) Define areas of ~0.5cm^2, as illustrated by the gray dotted line, and cut pieces for posterior longitudinal muscle dissection. (**b**) Flip the tissue to expose the serosa and longitudinal smooth muscle. At this stage, the expected orientation of the layers on the live imaging dish, from top to bottom, is described on the right. (**c**) Define a starting point (gray dots) and stop (orange dots) for dissecting the longitudinal muscle bundles, and continue dissecting until reaching the opposite side. From the new starting point (orange dotted line), repeat the same movement until a new starting point (green dotted line). Dissect to the bottom to generate the whole-mount circular muscle-myenteric plexus preparation. The dissection method described above is a suggestion; however, other methods of dissection achieve the same goals

6. Acclimate the cells to the optimal temperature and conditions for cellular processes (e.g., 37 °C/5%CO_2) (*see* **Note 5**).

3.3 Live Imaging Dish Setup

In this step, CMMP preparations were transferred to a microscope with an external gravity perfusion system setup (*see* **Note 6**). The perfusion system allows for a constant supply of warmed solutions. Warmed solutions from the perfusion system flow into the live imaging dishes through an inflow and exit using the outflow created by a peristaltic pump (*see* **Note 7**).

1. Fill the perfusion system compartments (e.g., syringes) with Krebs and desired treatment solutions such as agonists, antagonists, and blockers.
2. Turn on Krebs solution perfusion (2–3 mL/min was the inflow rate applied here).
3. Start Krebs solution perfusion for acclimatization for a few minutes.

4. Immerse the water-dipping objective lens in the Krebs solution where the tissue is located. With the help of the bright field, locate the depth at which the myenteric plexus is located. The myenteric plexus has a plexiform structure, with its ganglia interconnected with nerve fibers. Define and center the ganglion of interest to be studied.
5. If not already, move to the red channel to acquire a tdT picture. This picture will be used as a tool for region of interest identification using the animal model described here.
6. Move to the green channel to visualize the cells expressing the GECI GCaMP5.

3.4 Experimental Procedures

This method allows the evaluation of ENS-evoked responses by broad activation of myenteric ganglia.

3.4.1 Bath Application

1. Using the Krebs solution, record a baseline period before stimulation (~30 s).
2. After the baseline period, begin perfusing the desired treatment drugs by switching to the correct channel in the perfusion system.
3. Record until the evoked activity has ended (or following experimental goals).

The compartments of a perfusion system can contain different treatments, such as agonists and antagonists. The application time of these varies, depending on the nature of the target receptors/channels. Depending on the nature of individual experiments, the timing of stimulation and responses may vary because of variables such as the calcium indicator and speed of the perfusion system. For enteric glia stimulation, ADP is a well-documented driver of glial and neuronal calcium activity [21]. Timing and concentration will vary depending on experimental goals. Using ADP 100 μM stimuli for 30 s triggers strong enteric glia and neuron calcium responses.

Several aspects inherent to this activity can be quantified in the Fiji software.

3.4.2 Puffing Setup for Local Drug Application

1. Backfill the glass micropipette with the drug of interest.
2. Position the micropipette in the micromanipulator.
3. Move the micropipette using the micromanipulator (ROI 200, MPC 200) over the sample (*see* **Notes 8** and **9**).
4. Reach the target ganglion for stimulation. Raise the objective lens so that the micropipette is positioned below it. Position the micropipette close to the objective lens and find the end of the micropipette. Keep the micropipette tip in focus.

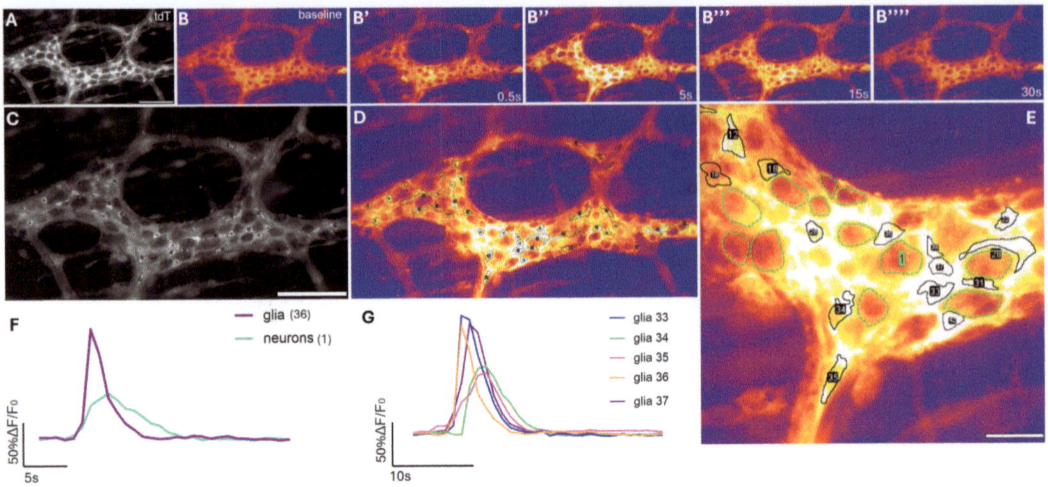

Fig. 3 Representative images of Ca^{2+} imaging. (**a**) High tdT reporter expression in enteric glia. (**b**) Baseline period before ADP bath application. (B′-B″) Evoked calcium response over time. (**c**) tdT picture with overlayed regions of interest (ROI) generated by the automatic detection method described in this chapter. Calibration bar: 50 μM. In (**d**), the peak of Ca^{2+} activity is with the glial ROIs. (**e**) Zoomed in on the ganglion with glial and neuron ROIs in the overlay. (**f**) Ca^{2+} traces of enteric glia (cell 36) and neuron (cell 1) activity over time due to ADP stimulation. Note that glia were recruited first, followed by enteric neuron activity. (**g**) Enteric glial cells'activities over time. (a–c) Calibration bar: 50 μM; (e) calibration bar: 25 μM

5. Ideally, advance the focus to find the tissue. Then, bring the micropipette into focus. Repeat these steps until to bring the tissue to focus.
6. Position the micropipette tip as close as possible to the target ganglion. A light touch on the muscle helps to check the depth of the micropipette and whether its position is optimal. This positioning must ensure that the drug application will be carried out on the target ganglion and that it will be washed due to the constant bath application (*see* **Note 11**).
7. Record a baseline period before local drug application.
8. Open the valve to deliver a volume of drug onto the enteric ganglion. Manual or automatic applications may apply.
9. Close the syringe valve to stop drug application.
10. Record until the evoked activity is ended (or following experimental goals) (Fig. 3a and b).

Multiple drugs have been used for enteric glial activation and verification of calcium activity, such as ADP (3 mM) and Lysophosphatidic acid (1 μM and 10 μM), while transgenic animals expressing DREADDs sensitive to N-clozapine in $GFAP^{+}$ cells are a specific tool to activate enteric glia only and follow calcium responses in the enteric network.

3.5 Calcium Imaging Parameters

1. To excite GCaMP5g, pass light through a 485/20-nm band-pass filter and filter the emission with a 515-nm long-pass filter.
2. For the tdTomato channel, excite by passing the light through a 535/20-nm band-pass filter and filter the reflected tdTomato (tdT) signals through a 610/75-nm band-pass emission filter before detection.
3. Image acquisition may be controlled by several commercial or open-source platforms, depending on availability and preference. In this example, Metamorph software was used.
4. Define acquisition speed. Based on the animal model used here, a GECI GCaMP5, the acquisition setup was 2 frames/s for 5 min (*see* **Note 10**).

3.6 Analysis: Basic Data Extraction Protocol Based on Cell Reporter and Automatic Detection of Regions of Interest

With the high expression of tdT reporter in enteric glia in the animal model reported here, the tdT image of the ganglion identifies enteric glial cells (Fig. 3a and c). Neurons are identified and delineated based on their morphology and/or evoked activity (Fig. 3c and e). Using Fiji software, tdT image processing includes the following steps:

3.6.1 tdT Image Processing

1. Open tdT image.
2. Define a region of interest (ROI) that includes the ganglion that was stimulated and click T. This defined ROI will be added to the ROI Manager.
3. ROI Manager: click More, Save. Later, this ROI will be applied to the recordings to define the quantification areas.
4. Then, click Image → Crop to obtain the image of the ganglion with the cropped ROI.
5. Click Process → FFT → Band-pass filter. The free Fourier transform (FFT) tool should have a setup to filter larger objects to 40 pixels and small objects to 3 pixels, with 5% directional tolerance permitted.
6. Click Adjust → Threshold. The resulting filtered image should be auto-thresholded using the Intermodes threshold algorithm.
7. The resulting binarized image should be analyzed: click Analyze → Analyze particles. The setup should include size: 400-pixel units in size to infinity and with the circularity of 0.0–1.0. The software is expected to create an ROI Manager window containing all ROIs based on tdT image processing. Otherwise, all ROIs should be manually added to the ROI Manager. In the ROI Manager, click More → Save.

3.6.2 Recordings Processing

1. The initially selected and saved ROI of the tdT image must be applied to the respective video, and this area cropped.
2. Apply ROI by dragging the file to Fiji, and it will automatically open in ROI Manager.
3. Apply your saved ROI over the video. Click Image → Crop (*see* **Note 12**). Several Fiji plugins, such as "BatchFolderProcessing" and "BatchVideoAlingment," can be applied to the recording to decrease background and movement.

3.6.3 Data Extraction

To determine which parameter will be extracted from your recordings, click Analyze → Set measurements → Mean Gray Value. Following are the steps to extract pixel values from your pre-determined ROIs:

1. Determine the baseline period and the activity start period (e.g., frames 1–60 as baseline; 61–400 activity).
2. Open the ROI file containing the glial ROIs automatically detected in Subheading 3.6.1.
3. Apply all glial ROIs (Fig. 3d). Align any if necessary by clicking "Show label" in the ROI Manager and moving the ROI.
4. In ROI Manager, click on More → Multimeasure. A result window will be generated with all frames and values for all your ROIs (e.g., cells) throughout the time.
5. For neuronal ROI generation, manually create neuronal ROIs using the cropped tdT or even the recording, by identifying and delineating neuronal cells (Fig. 3e). Then, follow **steps 4** and **5** detailed in Subheading 3.6.3.

3.6.4 Signal-to-Baseline Ratio

Create a signal-to-baseline ratio, also known as $\Delta F/F_0$.

This formula measures the variation of emitted fluorescence through time. F_0 represents the baseline fluorescence, while ΔF is the moment-by-moment variation from that baseline.

$$\Delta F/F0 = (F\text{–}F0)/F_0$$

Multiple analyses can be applied to study and compare calcium activity over time and population recruitment [11, 22, 23] (Fig. 3). To have representative videos, color ("Lookup tables") and temporal color codes (Image → Hyperstack → Temporal color code) can be applied to (Video 1, *see* Electronic Supplementary Material).

4 Notes

1. A careful and delicate tissue extraction and manipulation are essential to maintain the ENS structure and connection with its target cells (smooth muscle). Dispense a small volume of Krebs

solution into the newly opened abdominal cavity. This will help to separate the adhesion between the tissues and prevent the organs from drying out. A refresh of ice-cold Krebs solution throughout the protocol is required.

2. This cut, following the orientation of the mesenteric border line, guarantees the maintenance of the structure and orientation of the smooth muscle bundles, essential for microdissection.
3. The proportionality of tissue size will guarantee a good distribution of vector forces and stretching of the preparation. It is recommended to bring the dissection dishes to a vacuum chamber to eliminate any bubbles inside the Sylgard after a few uses. Inserting pins pushes air bubbles into the Sylgard, which can compromise calcium imaging acquisition. Mini insect-pins are custom cuts from a tungsten wire.
4. Refresh the ice-cold Krebs solution. Pull bundles to one point and then restart a new movement to another lower point, and so on. This will ensure that all longitudinal muscles will be microdissected, exposing the myenteric plexus. After longitudinal dissection, the tissue has likely moved or loosened. Repin the corners and where else you deem necessary so that the CMMP is well exposed and stretched.
5. The incubation time may vary, depending on the possible treatment or physiological manipulation that you wish to test on the calcium activity of cells that express GECI. In an example of a control situation, the live dish containing CMMP is incubated for 20–30 min at 37 °C/5%CO_2 before live imaging. Keep incubation and treatment times consistent between CMMP preparations to reduce variability during the execution of the experiment. In this sense, plan intervals to organize the next stage of the imaging setup and start calcium live imaging at the appropriate time.
6. Different kinds of microscope stages can be used (such as fixed or mechanical). This technique demands an upright microscope for the correct sample orientation for proper perfusion over the dishes to perform calcium imaging.
7. Test the perfusion system, inflow and outflow connections, outflow pump, and all other components that should work properly during live imaging. Commonly, perfusion systems get bubbles between uses, so constant flow in the perfusion system must be ensured. Preferably, test these steps with some advance.
8. This tool allows the positioning and movement of the micropipette over the CMMP preparation. The pipette micromanipulator is connected to a 1 mL syringe with a valve that can be manually or automatically manipulated to administer the

contents of the micropipette over the CMMP preparation. The manual application of the drug can be carried out through very slight manual positive pressure applied to the syringe apparatus, creating an application in a picoliter range. Another option is using an automatic pump with a defined volume and pressure. The local puffing technique is extremely delicate since the glass micropipette has a tip with a few μM openness. The movement of the filled micropipette must be as slow and delicate as possible, to ensure that during the positioning of the micropipette on the preparation, there is no mechanical stimulation, mechanical shock, or content leaking from the micropipette. Furthermore, the micropipette micromanipulator has different travel speeds and directions (left, right; forward and backward; up and down). These commands will guide the movement and positioning of the micropipette in the ganglion.

9. Depending on the quality of the micropipette (e.g., tip is not properly forged, aligned, symmetrical, or not broken), drug leak is possible. This can occur spontaneously without any need for pressure on the syringe or during local application to the ganglion. This demonstrates that there is a problem with the micropipette, and it must be replaced.
10. Several aspects of the calcium indicator will determine the acquisition speed, such as sensor signal strength, stability, transition kinetics, and molecule interactions [24]. Combining characteristics of the indicator with attributes of the biological events and cell types to be verified will determine optimal times for recording acquisition speed. The model used as an example here is GCaMP5, an indicator with a high dynamic range and high fluorescence emission after Ca^{2+} binding. GCaMP5 features such as lower Ca^{2+}-free fluorescence, high Ca^{2+}-bound fluorescence, and high Ca^{2+} affinity favor the acquisition of rapid events upon stimulation [19]. The calcium imaging recording setup (frames/s, total time) described here is 2 frames/s for 5 min. For a detailed evaluation of temporal responses, acquisition at different rates can be applied and will depend on the microscope features, software, etc.

References

1. Bootman MD (2012) Calcium Signaling. Cold Spring Harb Perspect Biol 4(7):a011171. https://doi.org/10.1101/cshperspect.a011171
2. Héja L, Nyitrai G, Kékesi O et al (2012) Astrocytes convert network excitation to tonic inhibition of neurons. BMC Biol 10(1):26. https://doi.org/10.1186/1741-7007-10-26
3. Kawamoto EM, Vivar C, Camandola S (2012) Physiology and pathology of calcium signaling in the brain. Front Pharmacol 3:61. https://doi.org/10.3389/fphar.2012.00061
4. Charles AC, Merrill JE, Dirksen ER et al (1991) Intercellular signaling in glial cells: calcium waves and oscillations in response to mechanical stimulation and glutamate. Neuron 6(6):983–992. https://doi.org/10.1016/0896-6273(91)90238-u
5. Cornell-Bell AH, Finkbeiner SM, Cooper MS, Smith SJ (1990) Glutamate induces calcium

waves in cultured astrocytes: long-range glial signaling. Science 247(4941):470–473. https://doi.org/10.1126/science.1967852

6. Corkrum M, Covelo A, Lines J et al (2020) Dopamine-evoked synaptic regulation in the nucleus accumbens requires astrocyte activity. Neuron 105(6):1036–1047.e5. https://doi.org/10.1016/j.neuron.2019.12.026
7. Shah D, Gsell W, Wahis J et al (2022) Astrocyte calcium dysfunction causes early network hyperactivity in Alzheimer's disease. Cell Rep 40(8):111280. https://doi.org/10.1016/j.celrep.2022.111280
8. Lia A, Sansevero G, Chiavegato A et al (2023) Rescue of astrocyte activity by the calcium sensor STIM1 restores long-term synaptic plasticity in female mice modelling Alzheimer's disease. Nat Commun 14(1):1590. https://doi.org/10.1038/s41467-023-37240-2
9. McClain JL, Fried DE, Gulbransen BD (2015) Agonist-evoked Ca(2+) signaling in enteric glia drives neural programs that regulate intestinal motility in mice. Cell Mol Gastroenterol Hepatol 1(6):631–645. https://doi.org/10.1016/j.jcmgh.2015.08.004
10. McClain J, Grubisic V, Fried D et al (2014) Ca2+ responses in enteric glia are mediated by connexin-43 hemichannels and modulate colonic transit in mice. Gastroenterology 146(2):497–507.e1. https://doi.org/10.1053/j.gastro.2013.10.061
11. Seguella L, McClain JL, Esposito G et al (2022) Functional intra- and inter-regional heterogeneity between myenteric glial cells of the colon and duodenum in mice. J Neurosci 42:JN-RM-2379-20. https://doi.org/10.1523/JNEUROSCI.2379-20.2022
12. Gulbransen BD, Sharkey KA (2009) Purinergic neuron-to-glia signaling in the enteric nervous system. Gastroenterology 136(4):1349–1358. https://doi.org/10.1053/j.gastro.2008.12.058
13. Grubisic V, Gulbransen BD (2017) Enteric glia: the most alimentary of all glia. J Physiol 595(2):557–570. https://doi.org/10.1113/JP271021
14. Scavuzzo MA, Letai KC, Maeno-Hikichi Y et al (2023) Enteric glial hub cells coordinate intestinal motility. bioRxiv [Preprint] 9: 2023.06.07.544052. https://doi.org/10.1101/2023.06.07.544052
15. Paredes RM, Etzler JC, Watts LT et al (2008) Chemical calcium indicators. Methods 46(3): 143–151. https://doi.org/10.1016/j.ymeth.2008.09.025
16. Lohr C, Beiersdorfer A, Fischer T et al (2021) Using genetically encoded calcium indicators to study astrocyte physiology: a field guide. Front Cell Neurosci 15:690147. https://doi.org/10.3389/fncel.2021.690147
17. Russell JT (2011) Imaging calcium signals in vivo: a powerful tool in physiology and pharmacology. Br J Pharmacol 163(8):1605–1625. https://doi.org/10.1111/j.1476-5381.2010.00988.x
18. Zhang Y, Rózsa M, Liang Y et al (2023) Fast and sensitive GCaMP calcium indicators for imaging neural populations. Nature 615(7954):884–891. https://doi.org/10.1038/s41586-023-05828-9
19. Akerboom J, Chen TW, Wardill TJ et al (2012) Optimization of a GCaMP calcium indicator for neural activity imaging. J Neurosci 32(40): 13819–13840. https://doi.org/10.1523/JNEUROSCI.2601-12.2012
20. Chen TW, Wardill TJ, Sun Y et al (2013) Ultrasensitive fluorescent proteins for imaging neuronal activity. Nature 499(7458):295–300. https://doi.org/10.1038/nature12354
21. Thomasi B, Gulbransen BD (2024) Su1217 glial S100B regulates enteric nervous system excitability in health and disease. Gastroenterology 166(5):S-697
22. Ahmadzai MM, Seguella L, Gulbransen BD (2021) Circuit-specific enteric glia regulate intestinal motor neurocircuits. Proc Natl Acad Sci USA 118(40):e2025938118. https://doi.org/10.1073/pnas.2025938118
23. Morales-Soto W, Gonzales J, Jackson WF (2023) Enteric glia promote visceral hypersensitivity during inflammation through intercellular signaling with gut nociceptors. Sci Signal 16(812):eadg1668. https://doi.org/10.1126/scisignal.adg1668
24. Zhong C, Schleifenbaum J (2019) Genetically encoded calcium indicators: a new tool in renal hypertension research. Front Med 6:128. https://doi.org/10.3389/fmed.2019.00128

Chapter 13

In Vitro Manipulation of Gene Expression in Enteric Glial Cells

Martina Vincenzi

Abstract

Emerging evidence suggests that enteric glial cells (EGCs) are important players in the intestinal tissue circuitry, where they control gut homeostasis, communicating with neurons and other neighboring cell types. However, the complex signalling pathways responsible for EGC functions are still largely unknown. To investigate the molecular mechanisms involved in cellular interactions and gut pathophysiology, EGC gene expression can be manipulated by transfection experiments. Here, we reported a protocol for gene knock-in and gene knock-down by transfection of EGCs with plasmid DNA and siRNA by lipofection.

Key words Enteric nervous system, Enteric glia, Transfection, Gene knock-in, Gene knock-down, Plasmid DNA, siRNA

1 Introduction

The enteric nervous system (ENS), also known as the "second brain," extends throughout the digestive tract, where it regulates multiple aspects of gastrointestinal physiology, including intestinal permeability, secretion, and motility [1–3]. The regulation of ENS functions depends on a complex bidirectional interplay between neurons and enteric glial cells (EGCs). Although the contribution of EGCs was long assumed to be merely supportive of the surrounding neurons, recent studies in glial biology have highlighted the heterogeneity, plasticity, and adaptability of EGCs, demonstrating their ability to undergo phenotypic and functional changes in response to various external signals [4–6]. These properties are critical for the dynamic interactions that EGCs have with neurons and other neighboring cells in the ganglia and throughout the intestinal wall, including epithelial, endocrine, and immune cells, to sustain local homeostasis [7–9]. Therefore, understanding the intricate communication pathways between these cell types is essential for deciphering the role of EGCs in health and disease.

Giuseppe Esposito (ed.), *Enteric Glial Cells: Methods and Protocols*, Methods in Molecular Biology, vol. 2971,
https://doi.org/10.1007/978-1-0716-4795-0_13,

Advances in this field could lead to improved treatments for a range of gastrointestinal disorders and highlight the overall importance of neuro-glial and immuno-glial interactions in the body.

In this context, manipulation of gene expression by transfection may allow scientists to better decipher the cellular interactions and molecular mechanisms involved in EGC-mediated control of intestinal pathophysiology. The following protocol describes the transfection of EGCs by lipofection. This method is optimized for the EGC line (EGC/PK060399egfr), an established, purified cell line isolated from the myenteric plexus of adult male rats and transformed with the egfr/neu vector containing the neomycin resistance gene and the gene encoding the intracellular c-neu tyrosine kinase and the extracellular receptor domains of the human epithelial growth factor, which are necessary to prevent infection and increase cell survival [10], which has been already used in several studies [11–14]. However, with further adjustments to the culture media, timing, and concentrations, this method can also be used for primary cultures of EGCs.

This protocol details all the procedures for achieving gene knock-in or knock-down in EGCs via transfection with plasmid deoxyribonucleic acid (DNA) or small interfering ribonucleic acid (siRNA), respectively. The outlined steps include cell culture (Subheading 3.1), coating of the cell culture device (Subheading 3.2), cell passaging and seeding (Subheading 3.3) and cell transfection (Subheading 3.4) (Fig. 1). Transfected cells obtained through this method can be utilized for investigating gene function and regulation, conducting mutational analysis and biochemical characterization of gene products, and assessing the effects of gene expression on the health and life cycle of EGCs alone and in the crosstalk with other cell types.

2 Materials

1. 1× Poly-L-lysine hydrobromide (Mw: 30,000–70,000) (*see* **Note 1**).
2. 1× sterile phosphate buffered saline (PBS) without Ca^{2+} and Mg^{2+}.
3. 37 °C water bath.
4. 50 mL conical tubes.
5. 70% Ethanol: Dilute absolute ethanol with distilled water.
6. Aceton.
7. Autoclaved fine straight forceps.
8. Centrifuge.

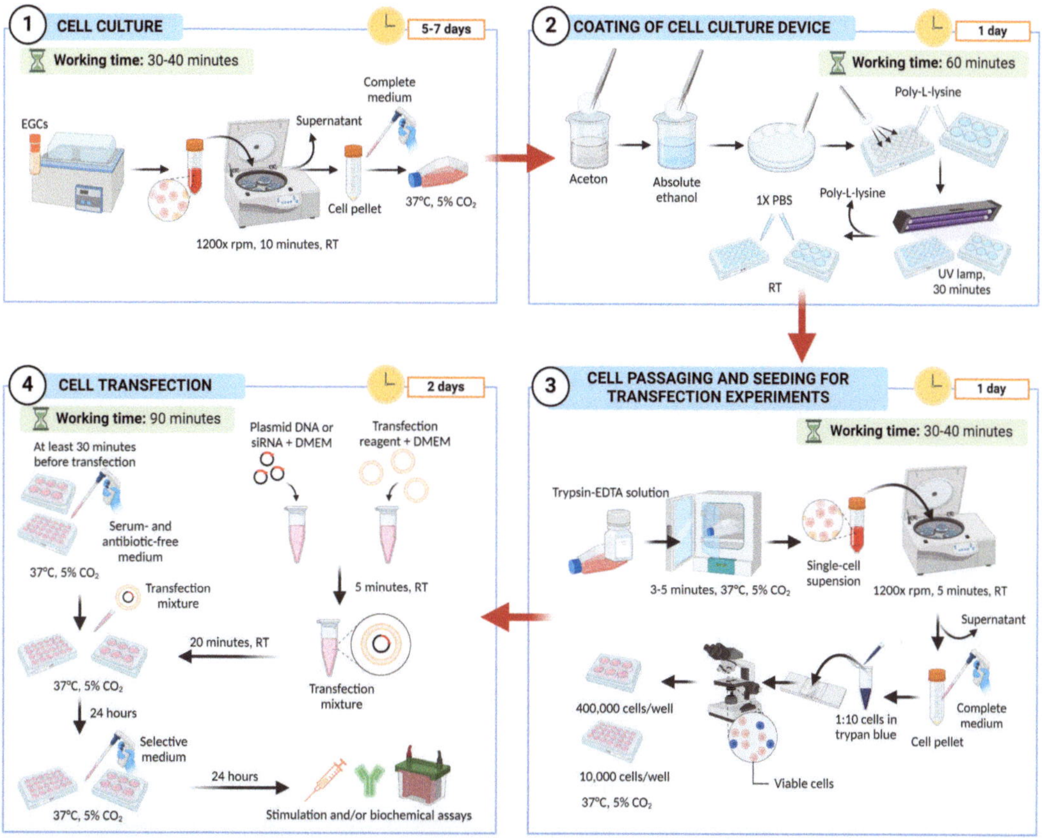

Fig. 1 Experimental scheme of enteric glial cells transfection

9. Complete medium: Dulbecco's Modified Eagle's Medium (DMEM), 10% fetal bovine serum (FBS), 2 mM L-glutamine, and 1% penicillin (100 U/mL)/streptomycin (100 μg/mL).
10. DMEM.
11. EGC/PK060399egfr cell line (EGC).
12. Ethanol, absolute alcohol without additive (purity ≥99.8%).
13. Incubator (37 °C, 5% CO_2, humidity atmosphere).
14. Laminar air flow hood equipped with UV lamp.
15. Light microscope.
16. Microscope cover glasses (13 mm and 20 × 20 mm).
17. Pipetman and sterile pipettes.
18. Pipettes and autoclaved tips (1000–200–10 μL).
19. Plasmid DNA of interest.
20. Selective medium: Dulbecco's Modified Eagle's Medium (DMEM), 10% fetal bovine serum (FBS), 2 mM L-glutamine, 1% penicillin (100 U/mL)/streptomycin (100 μg/mL), 0.2 mg/mL neomycin (G418).

21. Serum- and antibiotic-free medium: DMEM, 2 mM L-glutamine.
22. SiRNA of interest.
23. Sterile Petri dishes (100 × 25 mm).
24. Sterile plates (24-well and 6-well plates).
25. Syringes.
26. Syringe filters (0.22 μm).
27. T-75 flasks.
28. Transfection reagent such as Lipofectamine 2000 or DreamFect Gold.
29. Trypan blue.
30. Trypsin–EDTA solution.

3 Methods

3.1 Cell Culture

EGCs must be cultured 1 week prior to the transfection experiment to allow the cells to proliferate (*see* **Note 2**). From point 2, all subsequent steps must be carried out aseptically under hood with laminar airflow.

1. Thaw a new vial or a storage cryovial of EGCs quickly in a 37 °C water bath. Make sure that the lid does not fall into the water to avoid contamination.
2. After decontaminating the vial with 70% ethanol, transfer the contents to a 50 mL conical tube containing 10 mL of pre-warmed complete medium.
3. Centrifuge the cell suspension at 1200× rpm for 10 min at room temperature (RT) to remove the cryoprotective agent.
4. Discard the supernatant, resuspend the cell pellet by gently pipetting up and down in 10 mL of fresh complete medium, and distribute the cells in a T-75 flask.
5. Culture EGCs at 37 °C and 5% CO_2 and change the medium every two other days until 80% confluence is reached.

3.2 Coating of Cell Culture Devices

All steps in this section must be performed 1 day before EGC seeding under sterile conditions and under a laminar air flow hood. Both the microscope coverslips and the plates can be coated with poly-L-lysine, depending on the subsequent analysis (*see* **Note 3**). Please start with **step 1** of the following protocol for the coating of microscope coverslips and **step 4** for the coating of multi-well plates.

1. Quickly dip one microscope cover slip into acetone and then into absolute ethanol (*see* **Note 4**). Use fine, straight forceps to avoid contact between the gloves and coverslip.

2. Allow the microscope coverslips to dry. Use a sterile Petri dish to position the coverslips vertically and prevent them from sticking to the plastic surface when wet.
3. Place 13 mm round coverslips in 24-well plates and 20 × 20 mm square coverslips in 6-well plates. Select the most suitable device for your subsequent analysis.
4. Add 1 mL and 2.5 mL of poly-L-lysine to each well of the 24-well and 6-well plates, respectively (*see* **Note 5**).
5. After removing the lid, sterilize the plates under the UV light of the bonnet for 30 min.
6. Replace the poly-L-lysine with an equal amount of sterile 1× PBS and store the plates in the bonnet at RT until use.

3.3 Cell Passaging and Seeding for Transfection Experiments

All steps in this section must be performed under sterile conditions and under a laminar air flow hood.

1. Decant the complete medium from the sub-confluent culture vessel and rinse the cells with pre-warmed sterile 1× PBS to remove any traces of serum that may inhibit the action of the dissociation reagent.
2. After discarding the wash solution without disturbing the cell monolayer, add 3 mL of trypsin–EDTA reagent and incubate at 37 °C for 3–5 min.
3. Check the detachment of the cells under a light microscope. A gentle shaking may be required to completely detach the cells from the culture vessel.
4. After transferring the cell suspension to a 50 mL conical tube, use 5–6 mL of pre-warmed complete medium to detach the remaining cells, collect them in the same 50 mL conical tube, and inactivate the dissociation reagent.
5. Centrifuge the cell suspension at 1200× rpm for 5 min at RT.
6. After removing the supernatant, resuspend the cell pellet by carefully pipetting up and down in a small amount (3–5 mL) of fresh, pre-warmed complete medium.
7. Count the number of viable cells by a haemocytometer (*see* **Note 6**). Briefly, resuspend 100 μL of cell suspension in 900 μL of trypan blue (10×, dilution factor). Add 10 μL of trypan blue-cell suspension mixture in one chamber of the haemocytometer without moving the chamber coverslip, and allow the suspension to fill the chamber capillary. Count the viable cells (unstained cells) in the 1 × 1 mm central square and in two 1 × 1 mm corner squares using a 10× objective of a light microscope. The total number of cell/mL counted can be determined using the following calculation:

$$N = \overline{X} \text{ (average of cell counting in the 3 squares)} \times 10 \text{ (dilution factor)} \times 10^4 \text{ (chamber factor to calculate cells/mL)}$$

8. Dilute the cell suspension with the complete medium until the desired cell concentration is reached. For the transfection experiment, add 10,000 cells/0.5 mL of complete medium per well to 24-well plates and 400,000 cells/1.5 mL of complete medium per well to 6-well plates (*see* **Note 7**).
9. Incubate the plates at 37 °C, 5% CO_2 for 24 h.

3.4 Cell Transfection

All steps in this section must be performed under sterile conditions, in a laminar air flow hood.

1. Remove all medium from the multi-well plates and add 300 μL/well in 24-well plates and 1 mL/well in 6-well plates of pre-warmed serum- and antibiotic-free medium.
2. Incubate at 37 °C and 5% CO_2 in a humidified atmosphere for at least 30 min before transfection.
3. Prepare the transfection mixture as follows (all volumes indicated are for transfection of a single well. *See* **Note 8**).

   For 24-well plates:

   (a) Tube 1: Dilute a total amount of 500 ng of plasmid DNA and/or 300 ng of siRNA of interest in 50 μL of DMEM.

   (b) Tube 2: Dilute the transfection reagent (check the manufacturer's instructions for the volume and other details) in 50 μL of DMEM.

   For 6-well plates:

   (a) Tube 1: Dilute a total amount of 1000 ng of plasmid DNA and/or 600 ng of siRNA of interest in 150 μL of DMEM.

   (b) Tube 2: Dilute the transfection reagent (check the manufacturer's instructions for the volume and other details) in 150 μL of DMEM.
4. After 5 min of incubation at RT in the hood, combine the diluted plasmid DNA/siRNA dropwise with the diluted transfection reagent and incubate the transfection mixture for at least 20 min at RT under the hood to promote complex formation.
5. Slowly add 100 μL/well or 300 μL/well of the transfection complex mixture to 24-well or 6-well plates, respectively. Include control wells with cells only, cells with DNA only, cells with control siRNA from the same product line as the siRNA of interest, and cells with reagent only to account for any background effects.

6. Shake the plates carefully to ensure even distribution of the transfected material and return the plates to the incubator for 24 h.
7. Transfected EGC can be selected by replacing the serum- and antibiotic-free medium with the selective medium.
8. After 24 h of incubation at 37 °C and 5% CO_2, the transfected cells are ready for the biochemical analysis you are interested in. Perform immunofluorescence staining to check the efficiency of the transfection.

4 Notes

1. 1 mg/mL (100×) aliquots of poly-L-lysine hydrobromide (500 μL) can be prepared prior to the experiments by resuspending poly-L-lysine hydrobromide powder in sterile, double-distilled water and storing at −20 °C (long-term stability). The 100× aliquot can be quickly diluted in 50 mL of sterile, double-distilled water and filtered with 0.22 μm syringe filters before use (1× poly-L-lysine hydrobromide). 1× Poly-L-lysine can be replaced with 1× poly-D-lysine and supplemented with Laminin (50 μg/mL)
2. In general, it is recommended not to exceed the cell passage number 20 for transfection to avoid:
   (a) Genetic drift. With each passage, there is a risk that genetic changes or mutations accumulate in the cell population. These changes can alter the behavior and properties of the cells, which can affect the transfection efficiency and reproducibility of the experiments.
   (b) Phenotypic changes. A high number of passages can lead to changes in cell morphology, growth rates, and other phenotypic characteristics. This can affect the response of cells to transfection reagents and the uptake of genetic material.
   (c) Lower transfection efficiency. Cells with a higher passage number may have a lower transfection efficiency. This could be due to changes in cell membrane properties, changes in the cell cycle, or other factors that affect the cells' ability to take up and express transfected DNA or RNA.
   (d) Increased senescence. Cells may undergo replicative senescence after a certain number of passages, in which they stop dividing and show age-related changes. Senescent cells are usually less susceptible to transfection and other experimental manipulations.

(e) Reduced reproducibility. To ensure consistent and reproducible results, it is important to use cells with similar passage numbers. Using cells after the 20th passage can lead to fluctuations that make it difficult to interpret the data and compare different experiments.

3. Coated coverslips can be used for immunofluorescence staining whereas coated plates can be used for all other types of analysis, such as Western blots, crosslinking, and immunoprecipitation, that require a higher number of cells.
4. Aceton and ethanol are used to clean and degrease the surface of microscope cover glasses before use. This step can be replaced by autoclaving the coverslips.
5. The volume of poly-L-lysine per well can be reduced, i.e., 500 μL can be used for 24-well plates. Make sure that the volume of poly-L-lysine is sufficient to completely cover the coverslips.
6. Before starting, set the count parameters. It is not recommended to count cells touching the bottom and right double lines of the square chamber to not overestimating the cell number. If available, use an automatic cell counter.
7. The optimal number of cells/well varies depending on the cell line and status (growth rate, number of passages) and can therefore be changed. To avoid problems with transfection efficiency, plate the necessary cells to reach 70–90% confluence on the day of transfection.
8. Before starting the experiments, you may need to optimize transfection efficiency by systematically varying the amounts of reagents and plasmid DNA/siRNA you use in your transfection protocol. Although the common starting points are based on manufacturer recommendations, it is beneficial to try a broader range of concentrations for both the transfection reagent and the plasmid DNA/siRNA to determine the highest transfection efficiency with minimal cytotoxicity for your specific cell type and experimental setup.

   In our case, the optimal results are achieved with:

   (a) 500 ng plasmid DNA and/or 300 ng siRNA/50 μL of DMEM and 2 μL of Lipofectamine 2000 or 1.5 μL of DreamFect Gold/50 μL of DMEM for each well of 24-well plates.

   (b) 1000 ng plasmid DNA and/or 600 ng siRNA/150 μL of DMEM and 5 μL of Lipofectamine 2000 or 3 μL of DreamFect Gold/150 μL of DMEM for each well of 6-well plates.

References

1. Dockray G (1988) In: Furness JB, Costa M (eds) The enteric nervous system. ISBN 0443 032459 9. Price:£ 45.00. Churchill Livingston, Edinburgh, 1987. In Pergamon
2. Schneider S, Wright CM, Heuckeroth RO (2019) Unexpected roles for the second brain: enteric nervous system as master regulator of bowel function. Annu Rev Physiol 81: 235–259. https://doi.org/10.1146/annurev-physiol-021317-121515
3. Schemann M, Neunlist M (2004) The human enteric nervous system. Neurogastroenterol Motil 16(Suppl 1):55–59. https://doi.org/10.1111/j.1743-3150.2004.00476.x
4. Pochard C et al (2016) Defects in 15-HETE production and control of epithelial permeability by human enteric glial cells from patients with Crohn's disease. Gastroenterology 150(1):168–180. https://doi.org/10.1053/j.gastro.2015.09.038
5. Lefèvre MA, Soret R, Pilon N (2023) Harnessing the power of enteric glial cells' plasticity and multipotency for advancing regenerative medicine. Int J Mol Sci 24(15):12475
6. Santhosh S et al (2024) From diversity to disease: unravelling the role of enteric glial cells. Front Immunol 15:1408744. https://doi.org/10.3389/fimmu.2024.1408744
7. Aubé AC et al (2006) Changes in enteric neurone phenotype and intestinal functions in a transgenic mouse model of enteric glia disruption. Gut 55(5):630–637. https://doi.org/10.1136/gut.2005.067595
8. Van Landeghem L et al (2009) Regulation of intestinal epithelial cells transcriptome by enteric glial cells: impact on intestinal epithelial barrier functions. BMC Genomics 10(1):507. https://doi.org/10.1186/1471-2164-10-507
9. Progatzky F, Pachnis V (2022) The role of enteric glia in intestinal immunity. Curr Opin Immunol 77:102183. https://doi.org/10.1016/j.coi.2022.102183
10. Rühl A, Trotter J, Stremmel W (2001) Isolation of enteric glia and establishment of transformed enteroglial cell lines from the myenteric plexus of adult rat. Neurogastroenterol Motil 13(1):95–106. https://doi.org/10.1046/j.1365-2982.2001.00246.x
11. Costa DVS et al (2021) S100B inhibition attenuates intestinal damage and diarrhea severity during Clostridioides difficile infection by modulating inflammatory response. Front Cell Infect Microbiol 11:739874. https://doi.org/10.3389/fcimb.2021.739874
12. Macchioni L et al (2017) Enteric glial cells counteract Clostridium difficile toxin B through a NADPH oxidase/ROS/JNK/caspase-3 axis, without involving mitochondrial pathways. Sci Rep 7:45569. https://doi.org/10.1038/srep45569
13. Ghaisas S et al (2021) Chronic manganese exposure and the enteric nervous system: an in vitro and mouse in vivo study. Environ Health Perspect 129(8):87005. https://doi.org/10.1289/ehp7877
14. Kimono D et al (2019) Dysbiosis-associated enteric glial cell immune-activation and redox imbalance modulate tight junction protein expression in gulf war illness pathology. Front Physiol 10:1229. https://doi.org/10.3389/fphys.2019.01229

Chapter 14

In Vitro Study of Microbiota Influence on Enteric Glial Cell Physiology

Elena Lucarini and Alessandra Toti

Abstract

Gut microbiome is an essential partner in the maintenance of gastrointestinal (GI) homeostasis, through its interaction with both epithelium, immune cells, and enteric nervous system (ENS). Under pathological conditions, disturbances of resident bacteria composition and/or metabolism frequently combine with ENS dysfunction. Enteric glial cells (EGCs), accompanying neurons across all the layers of the GI tract, can orchestrate neuronal activity and immune response to microenvironment perturbation, but their interaction with the components of microbiota has been poorly investigated. Indeed, although a reciprocal influence between gut microflora and EGC has been suggested, further research is needed to elucidate the mechanisms governing this crosstalk and its implications in health and disease. To this end, in vitro studies appear necessary as they allow for dissection and selective analysis of microbiota-to-EGCs signaling without the intestinal contour. This chapter has the objective to describe a method to process fecal material from either animals or humans for in vitro studying microbiota-mediated modulation of EGCs activity. It also illustrates the advantage of using whole fecal supernatant (FS) over selected supernatant and/or living bacteria to in vitro mimic microbiota signaling and introduce the possibility to treat EGCs with the conditioned medium of FS pre-treated epithelial cells.

Key words Enteric glia, Gut microbiota, Fecal supernatant, Dysbiosis, Colon, Intestinal epithelium

1 Introduction

Gut microbiota defines a dynamic collection of microorganisms symbiotically living with the host, which is currently considered an external organ of the human body. Although both EGCs and microbiota actively participate in almost all GI functions [1–4], their reciprocal interaction has been poorly investigated. Kabouridis et al. demonstrated that EGCs colonize the intestinal mucosa after birth, in concomitance with the engraftment of gut microbiota, which critically regulates both postnatal settlement of EGCs in the intestinal mucosa and the ongoing supply of glia in adult mice [5]. It has been reported that, according to its composition and/or metabolism, microbiota can positively or negatively

Giuseppe Esposito (ed.), *Enteric Glial Cells: Methods and Protocols*, Methods in Molecular Biology, vol. 2971, https://doi.org/10.1007/978-1-0716-4795-0_14,

influence gut physiology and enteric glia functions. Increased levels of certain microbiota-derived metabolites, such as butyrate, can inhibit EGCs' proliferation in vitro [6], while secretome from *Bifidobacterium bifidum* and *Bacteroides fragilis* can modify the expression of neurotrophic factors, inflammasome partners, and cytokines in EGCs, contributing to the regulation of intestinal inflammation [7]. Although mucosal glia seems to be more likely influenced by microbes than intraganglionic glia [3], evidence to date suggests that the overall glial network can sense and respond to microbial molecules [8, 9]. Furthermore, the disruption of the intestinal barrier, as observed in certain pathological conditions [10], can facilitate microbes and their products' penetration within the deep intestinal layers and their contact with glia. Noteworthy, in vitro studies demonstrated the involvement of EGCs in the maintenance of the intestinal epithelial barrier [11] and their capacity to discriminate between pathogens and probiotics by modulating the expression of Toll-like receptors [9, 12]. Contrary to previous observations [13, 14], Vicentini et al. reported that antibiotics in mice affect the number of S100β^+ glial cells in the myenteric plexus of the ileum, but not in the colon, with no changes in the mucosal glia network [15], while Inlender et al. have advanced the hypothesis that homeostasis of mucosal glial cells in human gut is independent of microbiota [16]. These contradictory findings draw attention to the still poor knowledge on the homeostatic mechanisms regulating microbiota–glia interplay and its derangement under dysbiosis conditions, stressing the need for further investigations. This chapter describes a method to process fecal material from either animals or humans for in vitro studying the microbiota modulation of EGCs activity. Considering the interaction between different species and the influence of gut environment on microbes' signaling, the use of supernatant from selected bacteria as a stimulus for EGCs might represent a bias in terms of translational biology. The treatment with whole fecal supernatant instead permits to vehicle the entire secretome from microbes and host in vitro, reproducing either health or disease conditions. Moreover, removal of live bacteria during fecal material processing represents another technical advantage, as it allows the preservation of original content from lumen. Indeed, the large percentage of intestinal microbes lost under aerobic conditions (necessary for EGCs culture) would inevitably cause a shift in the composition of supernatant, decreasing the adherence between the in vitro model and the real in vivo state. It is also important to consider that differences in the type, distribution, and quantity of the microbial strains between intestinal regions [17–19] might generate different interactions with enteric glia phenotypes. Unlike animals, obtaining the lumen content of different intestinal region might be difficult in patients, narrowing the study of microbiota at the level of colon, which is anyway the most densely colonized part of the gut. Future research should focus on establishing whether microflora directly influence EGCs generation and migration, or instead if these processes are altered by the effects of microbiome on other cellular components of the intestinal environment, such as epithelial cells. In this regard, the protocol introduces the possibility to treat EGCs with the conditioned medium of FS pre-treated epithelial cells, which likely resembles the signaling pathway involved in microbiota–glia crosstalk in vivo [20]. Also, enteric glia might play a role in the maintenance of intestinal microflora equilibrium, as microbiota biodiversity has been found to positively correlate with glial-derived S100β protein [21, 22]. However, because of the complexity in reproducing glia phenotypes out of the intestinal context, the in vitro evaluation of glia secretome effects on microbes remains a technical challenge.

2 Materials

1. EGCs medium (according to the type of culture).
2. 1–5 mL syringes.
3. 1.5–5 mL tubes.
4. Lab spatulas.
5. Millex®-GS filter unit sterile 0.22 μm.
6. Bench-top refrigerated centrifuge.
7. Laminar flow hood.
8. −80 °C lab freezer.

3 Methods

Due to the impossibility of culturing bacteria in vitro in the same environment as EGCs, the use of whole FS containing microbiota secretome is a commonly accepted technique. The method described is based on mouse-derived fecal material, and the concentrations need to be adjusted according to different species from which the feces originate.

3.1 Obtaining Sterile FS

1. Collect 200 mg of feces from a pool of mice in a 1.5 mL tube (*see* **Note 1**).
2. Store several aliquots of feces at −80 °C.
3. Resuspend frozen feces in 1 mL of EGCs medium (according to the type of culture; *see* **Note 2**), obtaining a 200 mg/mL suspension (*see* **Note 3**).

4. Dissolve the feces with a spatula by applying a rotary motion from bottom to top to prevent the feces from compacting on the bottom (keep on ice throughout the procedure).
5. Centrifuge the suspension (2 min; 4 °C; 700× *g*).
6. The suspension is divided into two phases, a corpuscular one at the bottom and the supernatant, with a thin layer at the interface. Collect only the supernatant, avoiding taking the other components.
7. Under the hood, sterilize the solution by using an insulin syringe with a 0.22 μm filter (*see* **Note 4**).
8. Press the syringe downward firmly and repeatedly (about 60–80 times) to obtain 200–300 μL of sterile solution (keep on ice throughout the procedure).
9. FS needs to be used fresh and immediately once prepared, avoiding freezing and thawing (*see* **Note 5**).

A graphical illustration of the procedure has been reported in Fig. 1. The fresh FS stock solution (200 mg/mL) can be directly used to treat cells after its dilution in EGC medium. Before starting the experiment, it is necessary to perform a concentration-response

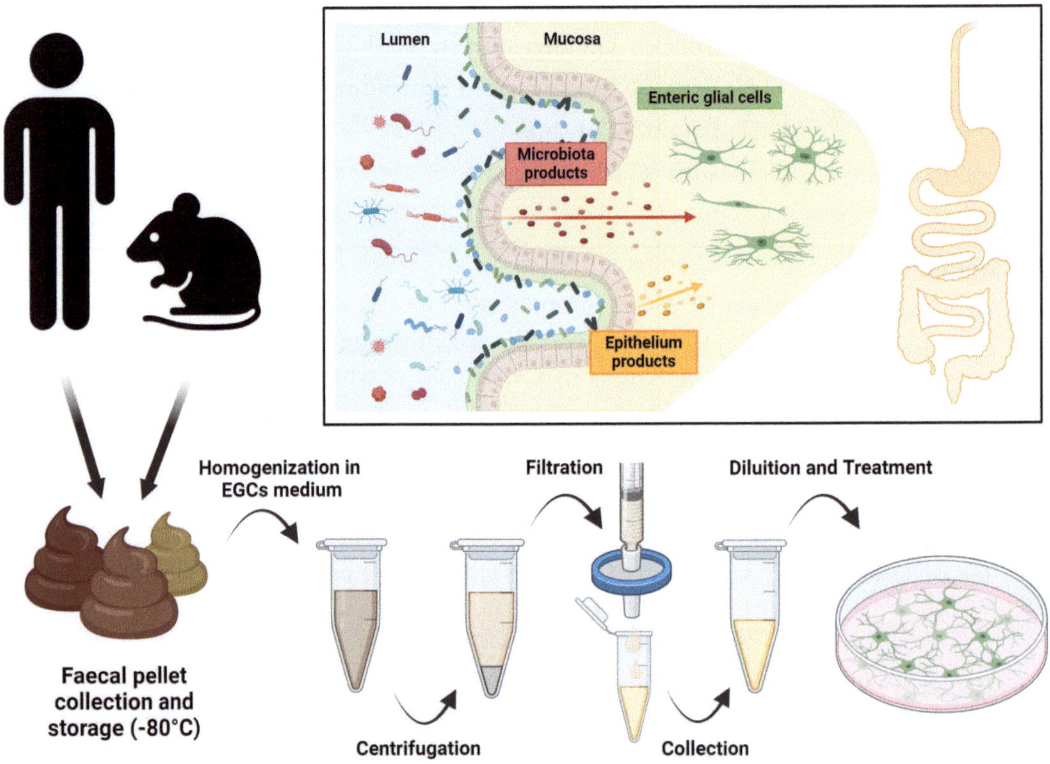

Fig. 1 Graphical illustration of fecal samples processing to obtain a suitable conditioned medium for the in vitro study of microbiota influence on EGCs

curve to identify a working solution that guarantees to provide cells with an adequate stimulus without causing toxicity.

Alternatively, a more complex protocol involving the treatment of EGCs with the conditioned medium from FS pre-treated epithelium cultures can be carried out.

3.2 Preparing Conditioned Medium from FS-Treated Epithelial Cells to Treat EGCs

1. FS can be used to treat various types of cultures replicating intestinal epithelium in vitro, i.e., Caco-2 cultures, Caco-2 and HT-29 co-cultures, or intestinal epithelial organoids (*see* **Note 6**).
2. After treating intestinal epithelial cultures with FS at the appropriate concentrations and time, the medium should be replaced with fresh medium for 24 h to allow FS-treated epithelium to release mediators (i.e., conditioned medium).
3. The conditioned medium of epithelial cells pre-treated with FS can be used to incubate EGC cells according to the type of analysis to be performed (*see* **Note 7**).

4 Notes

1. It is important to collect the appropriate amount of feces in the tube to resuspend it in the desired volume of medium and obtain the right stock concentration. This caution allows to avoid repeated sample thawing and freezing.
2. Resuspending the feces in the EGCs' culture medium overcomes the problem of further diluting FS for the treatment. Indeed, culture medium can be safely diluted with other buffers until a certain percentage, after which cell viability might be affected.
3. Various tests were conducted to determine the maximum filterable concentration of FS stock solution (200 mg/mL). The higher the concentration of the stock solution, the lower the volume of medium wasted at the time of treatment preparation, since the number of wells to treat can change between different experiments.
4. Numerous filtering tests were conducted sequentially using 70 μm, 40 μm, 0.45 μm, and 0.22 μm filters for obtaining sterile FS. However, it was observed that these preliminary steps resulted in material loss and did not facilitate the final filtration with the 0.22 μm filter.
5. It is important to apply a constant downward pressure on the entire syringe toward the filter with one hand and repeatedly press the plunger with the other. Indeed, the dense suspension can quickly clog the filter, thus causing its detachment from the syringe and fluid leaking.

6. It is important to ensure that intestinal epithelial cells and EGCs can be cultured in the same medium. Therefore, cell viability tests should be performed on EGCs incubated with the basal medium of epithelium cultures. If EGCs cannot be cultured in epithelial medium, it is recommended to test the compatibility of intestinal epithelial cell cultures in EGCs' basal medium.
7. It is suggested not to exceed 48 h of treatment, as it might impair the effectiveness of conditioned medium.

References

1. Cryan JF et al (2019) The microbiota-gut-brain axis. Physiol Rev 99(4):1877–2013. https://doi.org/10.1152/physrev.00018.2018
2. Sekirov I, Russell SL, Antunes LCM, Finlay BB (2010) Gut microbiota in health and disease. Physiol Rev 90(3):859–904. https://doi.org/10.1152/physrev.00045.2009
3. Seguella L, Gulbransen BD (2021) Enteric glial biology, intercellular signalling and roles in gastrointestinal disease. Nat Rev Gastroenterol Hepatol 18(8):571–587
4. Seguella L, Gulbransen BD (2021) Enteric glial biology, intercellular signalling and roles in gastrointestinal disease. Nat Rev Gastroenterol Hepatol 18(8):571–587. https://doi.org/10.1038/s41575-021-00423-7
5. Kabouridis PS et al (2015) Microbiota controls the homeostasis of glial cells in the gut lamina propria. Neuron 85(2):289–295. https://doi.org/10.1016/j.neuron.2014.12.037
6. Cossais F et al (2016) Postnatal development of the myenteric glial network and its modulation by butyrate. Am J Physiol-Gastrointest Liver Physiol 310(11):G941–G951. https://doi.org/10.1152/ajpgi.00232.2015
7. Yang P-c et al (2020) The influence of Bifidobacterium bifidum and Bacteroides fragilis on enteric glial cell–derived neurotrophic factors and inflammasome. Inflammation 43(6):2166–2177. https://doi.org/10.1007/s10753-020-01284-z
8. Rosenbaum C et al (2016) Activation of myenteric glia during acute inflammation in vitro and in vivo. PLoS One 11(3):e0151335. https://doi.org/10.1371/journal.pone.0151335
9. Turco F et al (2014) Enteroglial-derived S100B protein integrates bacteria-induced Toll-like receptor signalling in human enteric glial cells. Gut 63(1):105. https://doi.org/10.1136/gutjnl-2012-302090
10. Motta J-P et al (2021) Gastrointestinal biofilms in health and disease. Nat Rev Gastroenterol Hepatol 18(5):314–334. https://doi.org/10.1038/s41575-020-00397-y
11. Flamant M et al (2011) Enteric glia protect against Shigella flexneri invasion in intestinal epithelial cells: a role for S-nitrosoglutathione. Gut 60(4):473–484
12. Brun P et al (2015) Toll like receptor-2-regulates production of glial-derived neurotrophic factors in murine intestinal smooth muscle cells. Mol Cell Neurosci 68:24–35
13. De Vadder F et al (2018) Gut microbiota regulates maturation of the adult enteric nervous system via enteric serotonin networks. Proc Natl Acad Sci 115(25):6458–6463
14. Kabouridis Panagiotis S et al (2015) Microbiota controls the homeostasis of glial cells in the gut lamina propria. Neuron 85(2):289–295. https://doi.org/10.1016/j.neuron.2014.12.037
15. Vicentini FA et al (2021) Intestinal microbiota shapes gut physiology and regulates enteric neurons and glia. Microbiome 9(1):210. https://doi.org/10.1186/s40168-021-01165-z
16. Inlender T et al (2021) Homeostasis of mucosal glial cells in human gut is independent of microbiota. Sci Rep 11(1):12796. https://doi.org/10.1038/s41598-021-92384-9
17. Dieterich W, Schink M, Zopf Y (2018) Microbiota in the gastrointestinal tract. Med Sci (Basel) 6(4). https://doi.org/10.3390/medsci6040116
18. Zoetendal EG, Rajilić-Stojanović M, de Vos WM (2008) High-throughput diversity and functionality analysis of the gastrointestinal tract microbiota. Gut 57(11):1605–1615
19. Vuik F et al (2019) Composition of the mucosa-associated microbiota along the entire gastrointestinal tract of human individuals.

United European Gastroenterol J 7(7): 897–907
20. Margiotta F et al (2024) Investigating epithelial-neuronal signaling contribution in visceral pain through colon organoid-dorsal root ganglion neuron co-cultures. Neural Regen Res 19(6):1199–1200. https://doi.org/10.4103/1673-5374.386403
21. Romano Spica V et al (2023) S100B affects gut microbiota biodiversity. Int J Mol Sci 24(3): 2248
22. Orsini M et al (2020) In silico evaluation of putative S100B interacting proteins in healthy and IBD gut microbiota. Cells 9(7):1697

Chapter 15

Intracerebral Transplantation of Enteric Glia

Sirine Hacene, Nina Colitti, Isabelle Loubinoux, and Carla Cirillo

Abstract

Today, therapies for nerve tissue regeneration are limited by several factors, including the valid source of cells, the optimal administration route, and the reproducibility of the approach. Enteric glia, the glial population in the enteric nervous system, have advantageous properties which motivate an interest in considering them as promising candidates for neuro-regenerative medicine. Research in this direction needs to be performed in validated preclinical models of brain pathologies, of acute or neurodegenerative nature, where different cell candidates can be effectively tested. The common marmoset (*Callithrix jacchus*) responds to these needs. This non-human primate is very close to humans in terms of brain anatomy and function.

In this methodological chapter, we describe in detail the validity of the common marmoset as a preclinical model to test the safety of enteric glia transplantation. Specifically, we will present the method to amplify and characterize human enteric glia before transplantation and the stereotactic intracerebral injection of them into the perilesional brain regions in a preclinical model of ischemic stroke, whose validity for regenerative medicine strategies will be provided.

Key words Cell therapy, Enteric glia, Acute brain injury, Common marmoset, Tissue regeneration

1 Introduction

Following damage to nerve tissue, endogenous repair processes such as neurogenesis, gliogenesis, and angiogenesis are initiated [1]. However, these processes are limited, especially in the central nervous system. In this respect, cell therapy is emerging as a promising strategy for the regeneration of damaged nervous tissue [2–11]. Various cell sources have been tested in preclinical models of brain injury, including neural, embryonic, mesenchymal, and induced pluripotent stem cells. The use of both neural and embryonic stem cells is controversial due to their embryonic or fetal origin. Although easily amplified in culture [12], neural stem cells can generate tumors. Embryonic stem cells, on the other hand, may be subject to rejection after transplantation. Induced pluripotent stem cells offer an alternative to embryonic stem cells

Giuseppe Esposito (ed.), *Enteric Glial Cells: Methods and Protocols*, Methods in Molecular Biology, vol. 2971,
https://doi.org/10.1007/978-1-0716-4795-0_15,

[13]. Although they are attractive for autologous transplantation, preclinical studies have shown that their grafting into the mouse brain results in teratomas [14, 15]. Mesenchymal stem cells have paracrine activity, enabling them to secrete neurotrophic factors. This property, together with their low immunogenicity [16] and non-tumorigenicity, makes them promising candidates for transplantation. However, their persistence at the injection site is limited (approx. 5% survival) in the days following transplantation, thus limiting their efficacy [17]. To date, none of these cell sources has been clinically validated for the regeneration of damaged brain tissue.

A valid source of cells is the enteric nervous system, the most complex peripheral nervous system. Considered the "second brain" [18], the enteric nervous system contains glial cells like those of the brain (astrocytes, microglia, oligodendrocytes) [19], with neurogenic and neurotrophic potential. These cells are therefore ideal candidates for cell therapy in cerebral ischemia, combining cell replacement and tissue regeneration.

In this chapter, we present how enteric glia could be exploited via intracerebral transplantation in a preclinical model of brain injury. The latter is represented by the common marmoset (*Callithrix jacchus*). There are several advantages associated with the use of a non-human primate species, particularly in terms of phylogenetic proximity to humans (anatomical, structural, and behavioral similarities) [20] and ease of zootechnical management. Furthermore, for the brain injury model, previous studies carried out by the research team have established its validity [21, 22]. Here, we present the working protocol of the xenotransplantation of enteric glia in a marmoset model of brain injury. In this model, the structural and functional repercussions following brain injury can be estimated, using brain imaging (Magnetic Resonance Imaging) and sensorimotor evaluation (neurological scale and behavioral tests), respectively. Before performing cell transplantation, it is necessary to consider the feasibility of such a procedure. Thus, based on the data from the literature, the next sections will focus on the properties of enteric glia, enabling us to anticipate the benefits and potential limitations of their transplantation.

To date, studies of neurogenic enteric tissue transplants have been encouraging. These include those by Belkind Gerson et al. [23]. Their pioneering work confirmed the feasibility of neural stem cell transplantation of enteric neural stem cells (ENSCs) into brain parenchyma in rodent models. Local or systemic cell therapy showed that transplanted ENSCs survived, proliferated, and differentiated into neuronal and glial cells. In addition, transplanted cells can stimulate local endogenous neurogenesis significantly. It is therefore realistic to envisage comparable results for cell therapy for brain damage.

1.1 Marmoset: A Valid Non-human Primate Model to Test Cell-Based Therapy for Brain Injury

The common marmoset (*Callithrix jacchus*) is a small "New World" primate belonging to the Cebidae family and the species *Callithrix*. It is endemic to northern Brazil [20] and is highly adaptable to new environments (tropical Atlantic rainforest and thorny Caatinga forests) [20, 24]. It is a small primate, averaging between 14 and 18 cm. The average body weight in the wild is around 322 g, varying according to age and genus [25]. It is a diurnal, arboreal species whose individuals gather and live in social groups of 3–17 individuals in the wild [20]. Intraspecific communication is rich and largely achieved through visual, auditory, olfactory, and tactile signals. Criteria for diagnosing the common marmoset species include long white (3 cm) or grayish ear tufts, a white hairy patch on the forehead, a striated pattern on the caudal third of the back, and a light gray ringed pattern of light gray (narrow rings) and black (wide rings) on the tailcoat. This makes the marmoset an ecologically adaptive species. However, it also includes a wealth of social behaviors, which it is essential to recognize and study in captivity, particularly in the field of research.

In the experimental field, the marmoset proves to be an ideal compromise between rodents and large primates, in terms of both scientific reliability and ethical considerations [26]. First, as non-human primates (NHPs), the marmoset offers an advantage of greater phylogenetic proximity to humans than that of rodent models. The world of science also covers it for its great compliance. The marmoset is a small primate species, easy to restrain and handle. It also has a short gestation period, early sexual maturity, and a relatively high reproductive rate maintained in captivity [26]. In addition, the common marmoset displays a wide range of social behaviors, like those of humans [27]. This makes it easier to anticipate their technical management in captivity, on the one hand, and to exploit similar techniques on the other (e.g., behavioral assessment). In the field of neuroscience, a database on its neuroanatomy is currently available to scientists [27]. Like rodents, the marmoset is a lissencephalic species, i.e., its cerebral cortex is devoid of convolutions [28]. From a technical point of view, this particularity makes it an interesting laboratory model for studies at a cortical level (e.g., microelectrode implantation). The marmoset is also used in a wide variety of research projects. The marmoset is used, for example, in studies of the auditory [29] or visual [30] systems. Marmosets are also used as models in the study spinal cord [31]. Indeed, they are dexterous animals [28], adapted to studies of functional recovery of manual motor skills in the case of neurological lesions. What's more, marmosets show a capacity for learning and curiosity that facilitates training of iterative behaviors in the absence of dietary restriction [28]. Finally, they are increasingly included in research on aging, age-related diseases, and neurodegenerative diseases such as Parkinson's disease [32].

Marmosets are easy to manage, but only if precautions are taken when using them. Marmosets should be paired at least two at a time in a cage [33]. The groups are then made up of individuals of gender and social status compatible with peaceful cohabitation. This arrangement allows this social species to express an important part of its social behavior. In addition, it is preferable to place the marmoset cages side-by-side and/or facing each other, allowing each to communicate visually and vocally with other individuals [33]. As far as feeding is concerned, marmosets are provided with daily rations and water ad libitum [24]. Dietary restrictions should only be imposed when general anesthesia is necessary for surgery, imaging studies, and possibly behavioral assessment [34]. General health is monitored daily or twice daily by zootechnical staff. Clinical and, if necessary, paraclinical surveillance is carried out regularly by the laboratory's referring veterinarian.

Compared with rats, marmosets have a high white/gray matter ratio, close to that of humans, and a brain four times larger than that of the rat. In addition, there are many similarities between the motor systems of animals and the human motor system. One of them is the presence of the internal capsule. Walker et al. have demonstrated the general organization of motor cortical areas between marmosets, humans, and macaques [28], starting with the primary motor cortex (M1, area 4). Furthermore, despite inter-individual variation, marmosets show a somatotopic organization of M1 comparable to that of humans. The muscles of the pelvic limbs, axial limbs (neck, trunk, tail), thoracic limbs, and face are represented from the medial to the lateral side of M1. Finally, the fibers of the descending pathways linking M1 to the spinal motor neurons also undergo decussation. The movements of the muscles of a semicircle are therefore initiated by contralateral cortical territories (M1) [35]. Taken together, these characteristics, together with those of cerebral anatomy and functional anatomy, make the marmoset a suitable animal model for the exploration of cerebral motor deficits [27, 36].

2 Materials

2.1 Preparation for Surgery and Anesthesia (Commercial Drugs)

1. Oxytetracycline solution 200 mg/mL.
2. Buprenorphine solution 0.3 mg/mL.
3. Glycopyrrolate.
4. Dexamethasone solution 2 mg/mL.
5. Alfaxalone solution 10 mg/mL.
6. Isoflurane.
7. Pramocaïne chlorhydrate gel 1%.

2.2 Brain Injury

1. Pramocaïne chlorhydrate gel 1%.
2. Vetedine solution 10%.
3. Lidocaine solution 4 mg/kg.
4. Malonate 3 M solution: Diethyl malonate 99% in NaCl 0.9%.
5. Microliter syringe Hamilton 10 μL.
6. Resorbable thread Vicryl Rapid.
7. Dexamethasone solution 2 mg/mL.

2.3 Cell Culture

1. Dulbecco's Modified Eagle Medium/Nutrient F-12, 1:1 Mixture, 10% heat-inactivated fetal calf serum (FCS), 1× antibiotic–antimycotic solution (stock solution 100×).
2. Trypsin/EDTA solution 1× (Stock solution 10×).
3. Dulbecco's phosphate-buffered saline (DPBS) 1×.
4. 8-well slide culture chamber.

2.4 Cell Culture Fixation and Immunostaining

1. Paraformaldehyde solution: Paraformaldehyde powder 95% in Dulbecco's phosphate-buffered saline (DPBS) 1×.
2. Blocking buffer: Dulbecco's phosphate-buffered saline (DPBS) 1×, 0.1% Triton™ X-100 (laboratory grade), 4% Normal Goat Serum.
3. Mouse monoclonal anti-S100 beta antibody (4C4.9).
4. Rabbit polyclonal anti-GFAP antibody.

2.5 Intracardiac Perfusion and Tissue Fixation

1. Dolethal® solution 200 mg/mL.
2. Paraformaldehyde solution: Paraformaldehyde powder 95% in Dulbecco's phosphate-buffered saline (DPBS) 1×.
3. Sodium chloride 0.9% solution (292.2 g/L).
4. Sucrose solutions 10%, 20% and 30%: Sucrose powder (laboratory grade) in Dulbecco's phosphate-buffered saline (DPBS) 1×.
5. Microtome.

2.6 Brain Slice Immunostaining

1. Paraformaldehyde solution: Paraformaldehyde powder 95% in Dulbecco's phosphate-buffered saline (DPBS) 1×.
2. Blocking buffer: Dulbecco's phosphate-buffered saline (DPBS) 1×, 0.1% Triton™ X-100 (laboratory grade), 4% Normal Goat Serum.
3. Primary antibodies: Rabbit polyclonal anti-GFAP, Rabbit monoclonal Anti-AIF1/IBA1—clone (4S8I1), Rabbit polyclonal anti-NeuN, Goat polyclonal Anti-DCX.
4. Secondary antibodies: Goat anti-mouse IgG H&L (Alexa Fluor® 488), Goat anti-rabbit IgG H&L (Alexa Fluor®594).
5. Nuclei staining: DAPI (4′,6-diamidino-2-phenylindole).

3 Methods

The injury model described in this section has been validated and published, and it mimics ischemic stroke [21] via malonate injection. Malonate is a dicarboxylic acid compound that serves as a substrate for the enzyme malonyl-CoA synthetase (ACSF3), which synthesizes the coenzyme malonyl-CoA in mammalian mitochondrial metabolism [37]. This coenzyme is a major compound involved in the synthesis and oxidation of fatty acids. Malonate is a reversible, competitive inhibitor of the enzyme succinate dehydrogenase (SDH) [38, 39]. The latter is a mitochondrial enzyme involved in cellular respiration, synthesizing cellular energy in the form of ATP (adenosine triphosphate) [40]. By interfering with cellular respiration mechanisms, malonate exerts an energy deficit, leading to cytotoxic effects [14] and to neuronal and glial death [41]. The entire procedure includes a brain injury model; brain imaging using MRI; cell preparation and intracerebral transplantation; post-mortem histology.

3.1 Brain Injury Model

3.1.1 Preparation for Surgery

1. Animals need to be fasted for 12 h before surgery, with access to water maintained.
2. Two to three hours before anesthesia, antibiotic prophylaxis is initiated by intramuscular (IM) route with a long-acting antibiotic (oxytetracycline, 20 μg/kg).
3. Forty minutes before anesthesia, an analgesic (buprenorphine, 25 μg/kg) is injected IM to limit post-operative pain.
4. Ten minutes before anesthesia, an IM injection of glycopyrrolate (20 μg/kg) is given to limit ptyalism.
5. An anti-inflammatory treatment is administered intraperitoneally (IP) to limit cerebral edema (dexamethasone, 2 mg/kg).
6. The marmosets are then anesthetized with alfaxalone (7 mg/kg) given IM. Maintenance isoflurane (1–3%) associated oxygen saturation (SpO_2) is maintained above 95% and rectal temperature close to 37.5 °C (rectal probe coated with Pramocaïne chlorhydrate 1%), with the use of heat pads (*see* **Note 1**).

3.1.2 Brain Injury

The brain lesion is induced by an intracortical injection of malonate. The procedure is performed on common marmosets aged approximately 5 years and weighing approximately 390 g.

1. The marmosets are placed in a stereotaxic frame, in prone position, head held in place by ear bars covered with anesthetic cream (pramocaïne hydrochloride 1%).
2. Aseptic surgical conditions are applied to the scalp, the area of intervention, by applying a sterile sheet and cleaning the shaved head three times with alcohol (80%), followed by vetedine solution 10%.

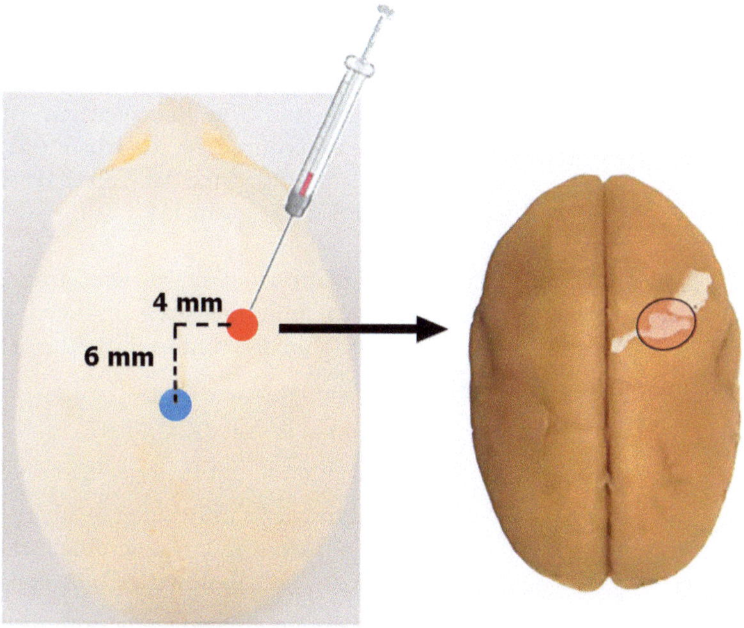

Fig. 1 Stereotactic localization of the intracerebral injection site. The Bregma (blue), point of intersection between the coronal and sagittal sutures, is the stereotactic landmark for determining the site of cranial puncture and intracortical injection (malonate). The drilling point (red) is located at the level of M1, controlling the motricity of the dominant thoracic limb (contralesional). (Adapted from [22])

3. Local anesthesia is provided by subcutaneous infiltration of lidocaine (4 mg/kg, 15 μL) at two injection points on either side of the future incision line.
4. The scalp is incised 10 min later. A trepanation hole (diameter 1 mm) is made, irrigated with 0.9% NaCl isotonic saline solution. The dura mater is then incised with a needle, providing access to the surface of the brain.
5. Intracortical injection of 8 μL of diethyl malonate at a concentration of 3 mol/L in 0.9% NaCl is initiated. Injection is performed using a 10 μL syringe (gauge: 0.485 mm, internal diameter: 0.22 mm) at an administration rate of 1 μL/min. Stereotactic coordinates are determined to target the primary motor cortex: cranio-caudal +6 mm, medio-lateral ±4 mm, and dorso-ventral −2 mm relative to the Bregma, according to the Stereotactic Atlas of the Marmoset Brain (Fig. 1) (*see* **Notes 2** and **3**).
6. The skin is then sutured using single-stitch, absorbable sutures resorbable thread (decimal 3.75 mm, round section).
7. Complete post-anesthetic recovery is carried out in the preparation room prior to return to the cage, where access to food

and water is guaranteed. Daily monitoring is carried out throughout the convalescence period. Post-operative recovery is observed on average 2–3 days after surgery.

8. Analgesia and the management of post-operative cerebral edema are ensured by corticosteroid therapy with dexamethasone (1 mg/kg, once daily, for 3 days, then 0.5 mg/kg, once a day for 3 days) given IM.

3.2 Brain Imaging

3.2.1 Choice of Parameters

MRI examinations of marmosets are carried out under general anesthetic to limit the stress induced by the procedure and to ensure the animals' immobility during image acquisition (*see* **Note 4**).

1. The animal is fasted for 12 h.
2. It is given an initial IM injection of alfaxalone (7 mg/kg).
3. The marmoset is positioned inside the antenna of the MRI, in a prone position under general anesthesia maintained with isoflurane at 1.5–3%, flow rate between 0.65 and 0.75 L/min.
4. Analysis of tissue damage, lesion volume, and edema over time is based primarily on T2-weighted MRI images.
5. At the end of the examination, the animals are returned to their living cage and monitored regularly until they wake up. Once fully awake, they are fed their usual ration.

3.2.2 Interpretation Methods

Analysis of tissue damage, lesion volume, and edema over time is based primarily on T2-weighted MRI image sequences. Lesion volume is measured by semi-automatic segmentation of regions of interest (ROIs), using MRIcron or ITKsnap software.

3.3 Preparation of Xenografts

3.3.1 Cell Culture

Enteric glia are isolated from human gut tissue, as described in detail in another chapter of this special issue, and following the protocol by Cirillo et al. [42].

1. Enteric glia are cultured in a Petri dish (10 cm) with cell culture medium, indicated in Subheading 2.4.
2. The plate is placed in the incubator (37 °C, 5% CO_2, and 95% relative humidity), and the medium is changed every 2 days.
3. A passage is performed every week for amplification: Once the medium is removed, the cells are washed with sterile PBS 1× and trypsinized (Trypsin/EDTA solution 1%) for 5 min at 37 °C. Cells are collected and placed in a 15 mL tube. After centrifugation (1400 rpm, 5 min, 4 °C), the supernatant is removed, and the pellet is resuspended in glial culture media.

3.3.2 Enteric Glia Characterization

Characterization of enteric glia cultures is performed using immunofluorescence. To this aim, the cells are placed in an 8-well slide culture chamber.

1. They are washed with DPBS before fixing the cells in paraformaldehyde (PFA) 4% in Dulbecco's phosphate-buffered saline (DPBS) 1×, for 30 min at room temperature.
2. Non-specific binding sites are masked with a blocking solution, indicated in Subheading 2.6, for 2 h at room temperature.
3. Cells are then incubated overnight at 4 °C in the presence of primary monoclonal mouse anti-S100β (1:1000) and polyclonal rabbit anti-GFAP (1:1000) antibodies.
4. The cultures are washed three times with DPBS 1× solution.
5. They are then incubated in the presence of a secondary antibody solution (goat anti-mouse 1:2000, and goat anti-rabbit 1: 2000 antibodies, respectively) and a blocking buffer at room temperature for 2 h (Fig. 2).
6. Negative controls ensure the absence of non-specific labeling. They are obtained by eliminating the incubation step in the presence of primary antibodies.

3.3.3 Xenograft Preparation Protocol

The xenograft is prepared from enteric glia detached from culture flasks by trypsinization.

1. The cells are washed with sterile DPBS 1× and trypsinized (Trypsin/EDTA solution 1%) for 5 min at 37 °C.

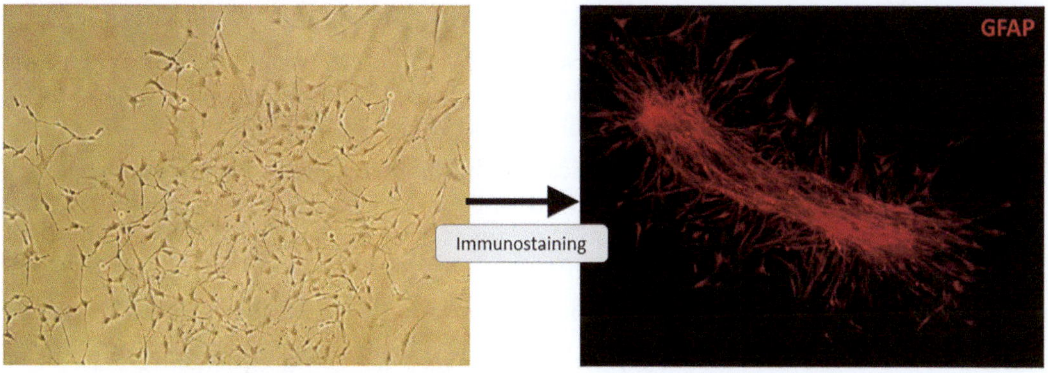

Human primary enteric glia.
Bright field image.

Human primary enteric glia.
Immunofluorescence with GFAP.

Fig. 2 Immunostaining of human primary enteric glia. Immunostaining is performed using antibodies directed against the glial cell marker GFAP. Immunohistochemical examination revealed positive labelling, confirming expression of the GFAP marker by the cells. Cell morphology under bright-field microscopy (left) and GFAP marker expression (right) demonstrate the glial phenotype of cultured cells. Observations using a ZEISS AXIOVERT 40 C microscope (3CCD camera) (left) and a ZEISS LSM710 confocal microscope (right). (Adapted from [2])

2. The cells are collected and placed in a 15 mL tube. After centrifugation (1400 rpm, 5 min, 4 °C), the supernatant is removed, and the pellet is resuspended in 1 mL of glial culture media.
3. The individualized cells are then counted, and a total of 1×10^6 enteric glia, distributed in 100 μL of glial growth medium, is prepared for the xenograft.

3.4 Transplantation

Prior to transplantation, the precise location and volume of the lesion are determined from T2 MRI images. These parameters determine the stereotactic coordinates of future transplant areas.

1. Animals need to be fasted for 12 h before surgery, with access to water maintained.
2. Two to three hours before anesthesia, antibiotic prophylaxis is initiated by intramuscular (IM) route with a long-acting antibiotic (oxytetracycline, 20 μg/kg).
3. Forty minutes before anesthesia, an analgesic (buprenorphine, 25 μg/kg) is injected IM to limit post-operative pain.
4. Ten minutes before anesthesia, a glycopyrrolate IM injection of glycopyrrolate (20 μg/kg) is given IM to limit ptyalism.
5. An anti-inflammatory treatment is administered intraperitoneal (IP) to limit cerebral edema (dexamethasone, 2 mg/kg).
6. The marmosets are then anesthetized with alfaxalone (7 mg/kg). Maintenance isoflurane (1–3%) associated oxygen saturation (SpO_2) is maintained above 95% and rectal temperature close to 37.5 °C (rectal probe coated with Pramocaïne chlorhydrate gel 1%), with the use of heat pads (*see* **Note 1**).
7. The marmosets are placed in a stereotaxic frame, in prone position, head held in place by ear bars covered with anesthetic gel (Pramocaïne chlorhydrate gel 1%).
8. Aseptic surgical conditions are applied to the scalp, the area of intervention, by applying a sterile drape and cleaning the shaved head three times with alcohol (80%), followed by vetedine solution 10%.
9. Local anesthesia is provided by subcutaneous infiltration of lidocaine (4 mg/kg, 15 μL) at two injection points on either side of the previous incision line (*see* Subheading 3.1.2).
10. Five to ten minutes later, the scalp is incised, and the galea (epicranial fascia connecting the eyelid muscles) is spread at the level of the previous surgical wound.
11. Cell transplantation is performed by distributing the total xenograft volume (100 μL) into four perilesional injection sites, via a Hamilton 25 mL syringe (series 701, gauge: 0.48 mm), through the same single hole, at a flow rate of between 7.3 and 8.3 μL/min (Fig. 3).

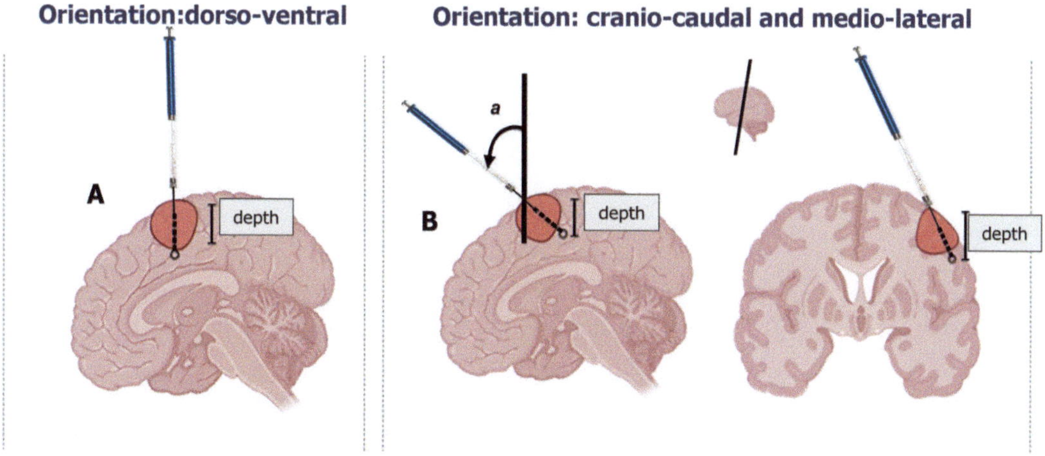

Fig. 3 Examples of the location of transplant sites and orientation of injections. Transplant sites can be determined by analysis of MRI images (T1, T2), to be located in the perilesional zone without damaging key cerebral structures (e.g., internal capsule, ventricles) or extracerebral structures. Tilt angles can be estimated from a vertical section

12. The syringe is left in place for an additional minute to limit cell reflux.
13. Cutaneous reconstruction is performed using single-stitch, resorbable sutures thread (decimal 3.0, 75 mm, round-section needle).
14. The animals are fully awakened in the preparation room before being returned to the cage access to food and water is re-established. Daily monitoring is carried out: General condition of coat, behavior, mobility, food intake, and body weight. Recovery from anesthesia is observed immediately and partial functional recovery 2–3 days post-postoperative.
15. Analgesia and the management of post-operative cerebral edema are ensured by corticosteroid therapy with dexamethasone (1 mg/kg, once daily, for 3 days, then 0.5 mg/kg, once a day for 3 days).

3.5 Intracardiac Perfusion and Brain Slice Preparation

At the end of the study, the marmoset will be sacrificed for brain analysis.

1. The animals are anesthetized with an IM injection of alfaxalone (9 mg/kg).
2. They are euthanized with an IP injection of sodium pentobarbital, 100 mg/kg.
3. After checking that there is no reflex (by pinching the interdigital space), the ribcage is opened using scissors.
4. The brain is then perfused with a buffer solution of warmed 0.9% NaCl/0.1% heparin buffer, followed by cooled PFA 4% administered intracardially.

5. A wide craniotomy, followed by extraction of the brain from the skull, is then performed.
6. The brain is placed overnight in a 4% PFA bath, followed by successive baths of sucrose solution in increasing concentrations (10%, 20% then 30%, each for 24 h).
7. Brains are cut into 30 μm coronal sections using a freezing microtome.

3.5.1 Immunostaining

Histological sections prepared in this way are arranged for staining and immunostaining.

1. The sections are mainly reserved for Nissl staining, to identify gray matter by staining of neuronal soma (Nissl bodies).
2. Other sections are stained with fast luxol blue (FLB) to highlight the white matter by staining the myelin.
3. A counterstain with cresyl violet to reveal the cytoarchitecture is performed every 12–14 coronal sections to locate the lesion site and adjacent structures.
4. Immunostaining can be performed to identify different cell types, including mature glial cells, reactive microglia, and neurons, using the markers S100β (1:1000), GFAP (1:1000), Iba1 (1:500), Doublecortin (1:500), and NeuN (1:500), respectively. To achieve this, sections are permeabilized and blocked by immersion in blocking buffer, indicated in Subheading 2.6, for 2 h at room temperature. The sections are then incubated for 24 h with primary antibodies (anti-S100β; anti-GFAP; anti-Iba1; anti-DCX; anti-NeuN) in blocking buffer solution. Following three successive washes with DPBS 1×, sections are incubated in the presence of the respective secondary antibodies (1:2000), in blocking buffer for 2 h at room temperature. In addition, cell nuclei are revealed by counterstaining with the marker DAPI (diluted 1:5000 in DPBS 1×). Images can be acquired with wide-field epifluorescence or confocal microscopes.

4 Notes

1. Vital parameters (heart rate, respiratory rate, temperature), in relation to signs of narcosis and analgesia, are regularly assessed throughout the surgical procedure. The usual values considered for marmoset species are as follows: heart rate assumed to be between 200 and 300 beats per minute (bpm); respiratory rate between 10 and 30 movements per minute (mpm) [43].
2. The cerebral hemisphere controlling the motricity of the dominant forelimb is targeted.

3. Following the injection, the needle is held in place for a further 5 min to limit any malonate reflux.
4. Each marmoset undergoes longitudinal evaluation by 24 h, 1 week, 4 weeks, and 12 weeks post-operatively, to evaluate brain injury (lesion volume, edema). As part of the clinical and paraclinical evaluation planned in the study, MRI scans can also be planned at different time points post-injury, i.e., at 6, 12, 18, and 24 months.

References

1. Cirillo C, Brihmat N, Castel-Lacanal E, Le Friec A et al (2020) Post-stroke remodeling processes in animal models and humans. J Cereb Blood Flow Metab 40(1):3–22. https://doi.org/10.1177/0271678X19882788
2. Hacene S, Le Friec A, Desmoulin F et al (2022) Present and future avenues of cell-based therapy for brain injury: the enteric nervous system as a potential cell source. Brain Pathol 32(5):e13105. https://doi.org/10.1111/bpa.13105
3. Rabinovich SS, Seledtsov VI, Banul NV et al (2005) Cell therapy of brain stroke. Bull Exp Biol Med 139(1):126–128. https://doi.org/10.1007/s10517-005-0229-y
4. Stem Cell Therapies as an Emerging Paradigm in Stroke Participants (2009) Stem cell therapies as an emerging paradigm in stroke (STEPS): bridging basic and clinical science for cellular and neurogenic factor therapy in treating stroke. Stroke 40(2):510–515. https://doi.org/10.1161/STROKEAHA.108.526863
5. Duncan T, Valenzuela M (2017) Alzheimer's disease, dementia, and stem cell therapy. Stem Cell Res Ther 8(1):111. https://doi.org/10.1186/s13287-017-0567-5
6. Genc B, Bozan HR, Genc S et al (2019) Stem cell therapy for multiple sclerosis. Adv Exp Med Biol 1084:145–174. https://doi.org/10.1007/5584_2018_247
7. Gómez RM, Sánchez MY, Portela-Lomba M et al (2018) Cell therapy for spinal cord injury with olfactory ensheathing glia cells (OECs). Glia 66(7):1267–1301. https://doi.org/10.1002/glia.23282
8. Assinck P, Duncan GJ, Hilton BJ et al (2017) Cell transplantation therapy for spinal cord injury. Nat Neurosci 20(5):637–647. https://doi.org/10.1038/nn.4541
9. Jovic D, Yu Y, Wang D et al (2022) A brief overview of global trends in MSC-based cell therapy. Stem Cell Rev Rep 18(5):1525–1545. https://doi.org/10.1007/s12015-022-10369-1
10. Kawabori M, Weintraub AH, Imai H et al (2021) Cell therapy for chronic TBI: interim analysis of the randomized controlled STEMTRA trial. Neurology 96(8):e1202–e1214. https://doi.org/10.1212/WNL.0000000000011450
11. Mangin G, Kubis N (2019) Cell therapy for ischemic stroke: how to turn a promising preclinical research into a successful clinical story. Stem Cell Rev Rep 15(2):176–193. https://doi.org/10.1007/s12015-018-9864-3
12. Ishibashi S, Sakaguchi M, Kuroiwa T et al (2004) Human neural stem/progenitor cells, expanded in long-term neurosphere culture, promote functional recovery after focal ischemia in Mongolian gerbils. J Neurosci Res 78(2):215–223. https://doi.org/10.1002/jnr.20246
13. Takahashi K, Tanabe K, Ohnuki M et al (2007) Induction of pluripotent stem cells from adult human fibroblasts by defined factors. Cell 131(5):861–872. https://doi.org/10.1016/j.cell.2007.11.019
14. Kawai H, Yamashita T, Ohta Y et al (2010) Tridermal tumorigenesis of induced pluripotent stem cells transplanted in ischemic brain. J Cereb Blood Flow Metab 30(8):1487–1493. https://doi.org/10.1038/jcbfm.2010.32
15. Miura K, Okada Y, Aoi T et al (2009) Variation in the safety of induced pluripotent stem cell lines. Nat Biotechnol 27(8):743–745. https://doi.org/10.1038/nbt.1554
16. Vu Q, Xie K, Eckert M et al (2014) Meta-analysis of preclinical studies of mesenchymal stromal cells for ischemic stroke. Neurology 82(14):1277–1286. https://doi.org/10.1212/WNL.0000000000000278
17. Amer MH, Rose FRAJ, Shakesheff KM et al (2017) Translational considerations in injectable cell-based therapeutics for neurological applications: concepts, progress and

challenges. NPJ Regen Med 10(2):23. https://doi.org/10.1038/s41536-017-0028-x
18. Furness JB (2012) The enteric nervous system and neurogastroenterology. Nat Rev Gastroenterol Hepatol 9(5):286–294. https://doi.org/10.1038/nrgastro.2012.32
19. Heanue TA, Pachnis V (2007) Enteric nervous system development and Hirschsprung's disease: advances in genetic and stem cell studies. Nat Rev Neurosci 8(6):466–479. https://doi.org/10.1038/nrn2137. Erratum in: Nat Rev Neurosci. 2007 Jul;8(7):568
20. De Fátima AM, Yamamoto ME, De Almeida Pessoa DM et al (2019) Taxonomy and natural history. In: The common Marmoset in captivity and biomedical research, pp 3–15
21. Le Friec A, Desmoulin F, Demain B et al (2021) Reproducible new model of focal ischemic injury in the marmoset monkey: MRI and behavioural follow-up. Transl Stroke Res 12(1):98–111. https://doi.org/10.1007/s12975-020-00804-1. Erratum in: Transl Stroke Res. 2021 12(1):204
22. Demain B. Preuve de concept d'une stratégie thérapeutique avec des neuro-implants microstructurés dans un nouveau modèle de lésion cérébrale focale chez le marmouset (Internet). 2015. Available from: https://tel.archives-ouvertes.fr/tel-01478801/.
23. Belkind-Gerson J, Hotta R, Whalen M et al (2016) Engraftment of enteric neural progenitor cells into the injured adult brain. BMC Neurosci 25(17):5. https://doi.org/10.1186/s12868-016-0238-y
24. Power ML, Koutsos L (2019) Marmoset nutrition and dietary husbandry. In: The common marmoset in captivity and biomedical research, pp 63–76
25. Tardif SD, Power ML, Ross CN et al (2013) Body mass growth in common marmosets: toward a model of pediatric obesity. Am J Phys Anthropol 150(1):21–28. https://doi.org/10.1002/ajpa.22110
26. Korte S, Everitt J (2019) The use of the marmoset in toxicity testing and nonclinical safety assessment studies. In: The common marmoset in captivity and biomedical research, pp 493–513
27. Fukushima M, Ichinohe N, Okano H (2019) Neuroanatomy of the marmoset. In: The common marmoset in captivity and biomedical research, pp 43–62
28. Walker J, MacLean J, Hatsopoulos NG (2017) The marmoset as a model system for studying voluntary motor control. Dev Neurobiol 77(3):273–285. https://doi.org/10.1002/dneu.22461
29. Eliades SJ, Tsunada J (2019) Marmosets in auditory research. In: The common marmoset in captivity and biomedical research, pp 451–475
30. Mitchell JF, Leopold DA (2015) The marmoset monkey as a model for visual neuroscience. Neurosci Res 93:20–46. https://doi.org/10.1016/j.neures.2015.01.008
31. Iwanami A, Yamane J, Katoh H et al (2005) Establishment of graded spinal cord injury model in a nonhuman primate: the common marmoset. J Neurosci Res 80(2):172–181. https://doi.org/10.1002/jnr.20435
32. Philippens IHCHM (2019) Marmosets in neurologic disease research: Parkinson's disease. In: The common marmoset in captivity and biomedical research, pp 415–435
33. Layne-Colon D, Goodroe A, Burns M (2018) Husbandry and housing of common marmosets. In: The common marmoset in captivity and biomedical research, pp 77–91. Available from: https://www.researchgate.net/publication/330056013_Husbandry_and_Housing_of_Common_Marmosets
34. Marini R, Wachtman L, Tardif S et al (2018) The common marmoset in captivity and biomedical research. In: The common marmoset in captivity and biomedical research, pp 1–570
35. Kondo T, Yoshihara Y, Yoshino-Saito K et al (2015) Histological and electrophysiological analysis of the corticospinal pathway to forelimb motoneurons in common marmosets. Neurosci Res 1(98):35–44. https://doi.org/10.1016/j.neures.2015.05.001
36. Van Essen DC, Donahue CJ, Coalson TS et al (2019) Cerebral cortical folding, parcellation, and connectivity in humans, nonhuman primates, and mice. Proc Natl Acad Sci USA 116(52):26173–26180. https://doi.org/10.1073/pnas.1902299116
37. Bowman CE, Rodriguez S, Selen Alpergin ES et al (2017) The mammalian malonyl-CoA synthetase ACSF3 is required for mitochondrial protein malonylation and metabolic efficiency. Cell Chem Biol 24(6):673–684.e4. https://doi.org/10.1016/j.chembiol.2017.04.009
38. Wojtovich AP, Smith CO, Haynes CM et al (2013) Physiological consequences of complex II inhibition for aging, disease, and the mKATP channel. Biochim Biophys Acta 1827(5):598–611. https://doi.org/10.1016/j.bbabio.2012.12.007
39. Henshaw R, Jenkins BG, Schulz JB et al (1994) Malonate produces striatal lesions by indirect NMDA receptor activation. Brain Res 647(1):161–166. https://doi.org/10.1016/0006-8993(94)91412-5

40. Pain F, Besret L, Vaufrey F et al (2002) In vivo quantification of localized neuronal activation and inhibition in the rat brain using a dedicated high temporal-resolution beta +-sensitive microprobe. Proc Natl Acad Sci USA 99(16): 10807–10812. https://doi.org/10.1073/pnas.162368899
41. Cuthill DJ, Fowler JH, McCulloch J et al (2006) Different patterns of axonal damage after intracerebral injection of malonate or AMPA. Exp Neurol 200(2):509–520. https://doi.org/10.1016/j.expneurol.2006.03.021
42. Cirillo C, Sarnelli G, Turco F et al (2011) Proinflammatory stimuli activates human-derived enteroglial cells and induces autocrine nitric oxide production. Neurogastroenterol Motil 23(9):10–12. https://doi.org/10.1111/j.1365-2982.2011.01748.x
43. Kramer R, Burns M (2019) Normal clinical and biological parameters of the common marmoset (Callithrix jacchus). In: The common marmoset in captivity and biomedical research, pp 93–107

Index

A

C

D

E

F

G

H

I

Giuseppe Esposito (ed.), *Enteric Glial Cells: Methods and Protocols*, Methods in Molecular Biology, vol. 2971, https://doi.org/10.1007/978-1-0716-4795-0,

MIX
Papier aus verantwortungsvollen Quellen
Paper from responsible sources
FSC® C105338

If you have any concerns about our products,
you can contact us on
ProductSafety@springernature.com

In case Publisher is established outside the EU,
the EU authorized representative is:
Springer Nature Customer Service Center GmbH
Europaplatz 3, 69115 Heidelberg, Germany

Printed by Libri Plureos GmbH
in Hamburg, Germany